Firefighter Fatalities
in the United States in 2003

U.S. Department of Homeland Security

Federal Emergency Management Agency

U.S. Fire Administration

August 2004

In memory of all firefighters
who answered their last call in 2003

To their families and friends

To their service and sacrifice

U.S. Fire Administration
Mission Statement

As an entity of the Federal Emergency Management Agency, the mission of the United States Fire Administration is to reduce life and economic losses due to fire and related emergencies, through leadership, advocacy, coordination, and support. We serve the Nation independently, in coordination with other Federal agencies, and in partnership with fire protection and emergency service communities. With a commitment to excellence, we provide public education, training, technology, and data initiatives.

On March 1, 2003, FEMA became part of the U.S. Department of Homeland Security. FEMA's continuing mission within the new department is to lead the effort to prepare the Nation for all hazards and effectively manage Federal response and recovery efforts following any national incident. FEMA also initiates proactive mitigation activities, trains first responders, and manages Citizen Corps, the National Flood Insurance Program, and the U.S. Fire Administration.

ACKNOWLEDGMENTS

This study of firefighter fatalities would not have been possible without the cooperation and assistance of many members of the fire service across the United States. Members of individual fire departments, chief fire officers, United States Forest Service personnel, the Department of Justice, the National Fire Protection Association (NFPA), the National Fallen Firefighters Foundation, and many others contributed important information for this report.

C² Technologies, Inc., of Vienna, Virginia, conducted this analysis for the United States Fire Administration (USFA) under contract EME-2003-CO-0282.

The ultimate objective of this effort is to reduce the number of firefighter deaths through an increased awareness and understanding of their causes and how they can be prevented. Firefighting, rescue, and other types of emergency operations are essential activities in an inherently dangerous profession, and unfortunate tragedies do occur. This is the risk all firefighters accept every time they respond to an emergency incident. However, the risk can be greatly reduced through efforts to improve training and operations at emergency scenes, and by increasing firefighter health and safety.

Photographic Acknowledgments

The USFA would like to extend its thanks to the following individuals for providing photographs for this report:

Front Cover Image: Rick Cinclair, Worcester Telegram & Gazette
A firefighter braces against blowing snow at the funeral for Firefighter Martin H. McNamara, V.

Betty Jenewin, Worcester Telegram & Gazette
Lancaster, Massachusetts, firefighter Martin H. McNamara, V, was killed while he fought a fire in the basement of this house. 14

Ohio State Highway Patrol
Scene of the crash that claimed the life of Firefighter Richard A. Long of the Gallipolis (Ohio) Volunteer Fire Department. The apparatus left the right side of the roadway, came back onto the road, and rolled. 22

Dave Schwarz, St. Cloud Times
St. Cloud, Minnesota, Assistant Chief Don Billig was killed when he was struck by a pickup truck at the scene of an emergency. 23

Glenn Hartong, Cincinnati Enquirer
Cincinnati Fire Chief Robert Wright carries the flag that covered Firefighter Oscar Armstrong's casket as he prepares to present it to Armstrong's mother. 62

Table of Contents

BACKGROUND

For 27 years, the United States Fire Administration (USFA) has tracked the number of firefighter fatalities and conducted an annual analysis. Through the collection of information on the causes of firefighter deaths, the USFA is able to focus on specific problems and direct efforts toward finding solutions to reduce the number of firefighter fatalities in the future. This information also is used to measure the effectiveness of current programs directed toward firefighter health and safety.

One of the USFA's main program goals is a 25-percent reduction in firefighter fatalities in 5 years, and a 50-percent reduction within 10 years. The emphasis placed on these goals by the USFA is underscored by the fact that these goals represent one of the five major objectives that guide the actions of the USFA.

In addition to the analysis, the USFA provides a list of firefighter fatalities to the National Fallen Firefighters Foundation. If Memorial criteria are met, the fallen firefighter's next of kin, as well as members of the individual fire department, are invited to the annual Fallen Firefighters Memorial Service. The service is held at the National Emergency Training Center in Emmitsburg, Maryland, during Fire Prevention Week. Additional information regarding the Memorial Service can be found at www.firehero.org or by calling the National Fallen Firefighters Foundation at (301) 447-1365.

Other resources and information regarding firefighter fatalities, including current fatality notices, the National Fallen Firefighters Memorial database, and links to the Public Safety Officer's Benefit (PSOB) program can be found at www.usfa.fema.gov.

INTRODUCTION

This report continues a series of annual studies by the USFA of on-duty firefighter fatalities in the United States.

The specific objective of this study is to identify all on-duty firefighter fatalities that occurred in the United States and its protectorates in 2003 and to analyze the circumstances surrounding each occurrence. The study is intended to help identify approaches that could reduce the number of firefighter deaths in future years.

In addition to the 2003 overall findings, this study includes two special topics related to alcohol use and fire service risk management.

Who is a Firefighter?

For the purpose of this study, the term *firefighter* covers all members of organized fire departments in all States, the District of Columbia, and the Territories of Puerto Rico, the Virgin Islands, American Samoa, the Commonwealth of the Northern Mariana Islands, and Guam. It includes career and volunteer firefighters; full-time public safety officers acting as firefighters; State, Territory, and Federal government fire service personnel, including wildland firefighters; and privately employed firefighters, including employees of contract fire departments and trained members of industrial fire brigades, whether full- or part-time. It also includes contract personnel working as firefighters or assigned to work in direct support of fire service organizations.

Under this definition, the study includes not only local and municipal firefighters but also seasonal and full-time employees of the United States Forest Service, the Bureau of Land Management, the Bureau of Indian Affairs, the Bureau of Fish and Wildlife, the National Park Service, and State wildland agencies. The definition also includes prison inmates serving on firefighting crews; firefighters employed by other governmental agencies, such as the United States Department of Energy; military personnel performing assigned fire suppression activities; and civilian firefighters working at military installations.

What Constitutes an On-Duty Fatality?

On-duty fatalities include any injury or illness sustained while on duty that proves fatal. The term *on-duty* refers to being involved in operations at the scene of an emergency, whether it is a fire or non fire incident; responding to or returning from an incident; performing other officially assigned duties such as training, maintenance, public education, inspection, investigations, court testimony, and fundraising; and being on-call, under orders, or on standby duty except at the individual's home or place of business. An individual who experiences a heart attack or other fatal injury at home as he or she prepares to respond to an emergency is considered on duty when the response begins. A firefighter who becomes ill while performing fire department duties and suffers a heart attack shortly after arriving home or at another location may be considered on duty since the inception of the heart attack occurred while the firefighter was on duty.

On December 15, 2003, the President of the United States signed into law the Hometown Heroes Survivors Benefit Act of 2003. After being signed by the President, the Act became Public Law 108-182. The law presumes that a heart attack or stroke are in the line of duty if the firefighter was engaged in nonroutine stressful or strenuous physical activity while on duty and the firefighter becomes ill while on duty or within 24 hours after engaging in such activity. The full text of the law is available at: http://frwebgate.access.gpo.gov/cgi-bin/getdoc.cgi?dbname=108_cong_public_laws&docid=f:publ182.108.pdf.

The inclusion criteria for this study will be affected by this change in the law. Previous to December 15, 2003, firefighters who became ill as the result of a heart attack or stroke after going off duty needed to register some complaint of not feeling well while still on duty in order to be included in this study. For firefighter fatalities after December 15, 2003, firefighters will be included in this study if they become ill as the result of a heart attack or stroke within 24 hours of a training activity or emergency response. Firefighters who become ill after going off duty where the activities while on duty were limited to nonstressful tasks that did not involve physical exertion such as clerical, administrative, or nonmanual in nature, will not be included in this study.

A fatality may be caused directly by an accidental or intentional injury in either emergency or nonemergency circumstances, or it may be attributed to an occupationally-related fatal illness. A common example of a fatal illness incurred on duty is a heart attack. Fatalities attributed to occupational illnesses also would include a communicable disease contracted while on duty that proved fatal when the disease could be attributed to a documented occupational exposure.

Injuries and illnesses are included even when death is considerably delayed after the original incident. When the incident and the death occur in different years, the analysis counts the fatality as having occurred in the year in which the incident took place.

For example, three firefighters died in 2003 as the result of injuries or exposures that they received while on duty previous to 2003. The USFA was notified of the death of one firefighter in 2001 that was not known or included in the firefighter fatality report for that year.

Information about these four deaths is included in the appendix of this report, but they are not addressed in the body of the report unless the death affects retrospective statistical comparisons.

There is no established mechanism for identifying fatalities that result from illnesses such as cancer that develop over long periods of time, which may be related to occupational exposure to hazardous materials or products of combustion. It has proved to be very difficult over the years to provide a complete evaluation of an occupational illness as a causal factor in firefighter deaths due to the following limitations: insufficient tracking of the exposure of firefighters to toxic hazards, the often delayed long-term effects of such toxic hazard exposures, and the exposures firefighters may receive while off duty.

Sources of Initial Notification

As an integral part of its ongoing program to collect and analyze fire data, USFA solicits information on firefighter fatalities directly from the fire service and from a wide range of other sources. These sources include the Public Safety Officer's Benefit (PSOB) program administered by the Department of Justice, the National Institute for Occupational Safety and Health (NIOSH), the Occupational Safety and Health Administration (OSHA), the United States military, the National Interagency Fire Center, and other Federal agencies.

The USFA receives notification of some deaths directly from fire departments, as well as from such fire service organizations as the International Association of Fire Chiefs (IAFC), the International Association of Fire Fighters (IAFF), NFPA, the National Volunteer Fire Council (NVFC), State fire marshals, State training organizations, other State and local organizations, fire service Internet sites, news services, and fire service publications. The USFA also keeps track of fatal fire incidents as part of its Major Fires Investigation Program and performs an ongoing analysis of data from the National Fire Incident Reporting System (NFIRS).

Procedure for Including a Fatality in the Study

In most cases, after notification of a fatal incident, initial telephone contact is made with local authorities by the USFA to verify the incident, its location, jurisdiction, and the fire department or agency involved. Further information about the deceased firefighter and the incident may be obtained from the chief of the fire department or his or her designee over the phone or by other data collection forms.

Information that is requested routinely includes NFIRS-1 (incident) and NFIRS-3 (fire service casualty) reports, the fire department's own incident reports and internal investigation reports, copies of death certificates or autopsy results, special investigative reports, police reports, photographs and diagrams, and newspaper or media accounts of the incident. Information on the incident may also be gathered from NFPA or NIOSH reports on an incident.

After obtaining this information, a determination is made as to whether the death qualifies as an on-duty firefighter fatality according to the previously described criteria. With the exception of firefighter deaths after December 15, 2003, the same criteria were used for this study as in previous annual studies. Additional information may be requested, either by followup with the fire department directly, from State vital records offices, or from other agencies. The determination as to whether a fatality qualifies as an on-duty death for inclusion in this statistical analysis is made by the USFA. The final determination as to whether a fatality qualifies as a line-of-duty death for inclusion in the Fallen Firefighters Memorial Service is made by the National Fallen Firefighters Foundation.

2003 FINDINGS

One hundred and eleven firefighters died while on duty in 2003. This level of fatalities continues a disturbing upward trend in the number of firefighter fatalities. Even if the horrible toll of September 11, 2001, is set aside momentarily, more firefighters are dying each year on duty in the last several years than would be expected after the lower loss years in the 1990's.

Figure 1. On-Duty Firefighter Fatalities (1977-2003)

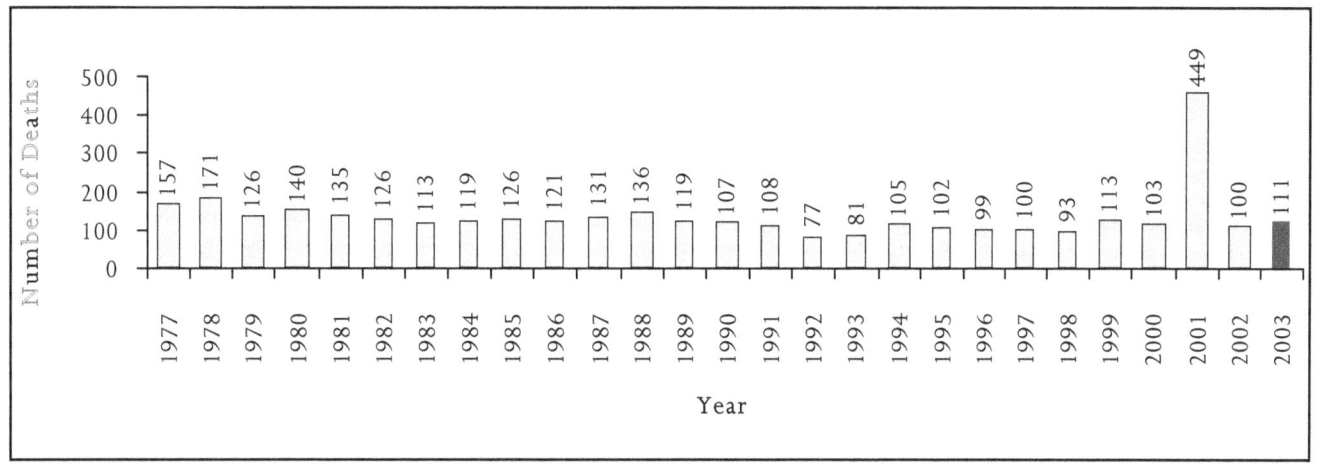

With the deaths of 111 firefighters in 2003, this is the eighth time in the past 10 years, and the eleventh time within the past 15 years, when the total number of firefighter fatalities has reached or exceeded 100. The lowest years on record are 1992 with 77 fatalities and 1993 with 81 fatalities (Figure 1).

The 111 deaths in 2003 are 111 percent of the 10-year average and 108 percent of the 5-year average. The 111 deaths resulted from a total of 98 incidents. There were seven multiple firefighter fatality incidents.

In 2001, 344 firefighters were killed as a result of the attacks on the World Trade Center (WTC) in New York City on September 11, 2001. When conducting multiyear comparisons of firefighter fatalities in this report, it may be necessary to set these deaths apart for illustrative purposes. This action is by no means a minimization of the supreme sacrifice made by these firefighters.

> **Seven Firefighters were murdered in 2003 while responding to or working at the scene of fires started by arsonists.**
>
> This total includes two Memphis firefighters who died in a commercial structure fire started by an employee to hide a crime.

While the total number of firefighter fatalities has been trending downward over the past 20 years, the number of firefighter deaths per fire incident has actually risen.

> The median age for firefighters who died while on duty in 2003 was 46 years and 5 months. Two 16-year-old firefighters died in 2003 and the oldest firefighter to die was 81 years of age.

7

The chart below (Figure 2) compares the total number of firefighter fatalities each year that are associated with responses to fires and the total number of fire incidents reported by NFPA International through 2002 (2003 data is not yet available). Despite a downward dip in the early 1990's, the level of firefighter fatalities is back up to the same levels experienced in the 1980's. If the firefighter deaths at the WTC are included in the 2001 data, the number rises to 23.1 firefighter fatalities per 100,000 fires.

Figure 2. Firefighter Fatalities per 100,000

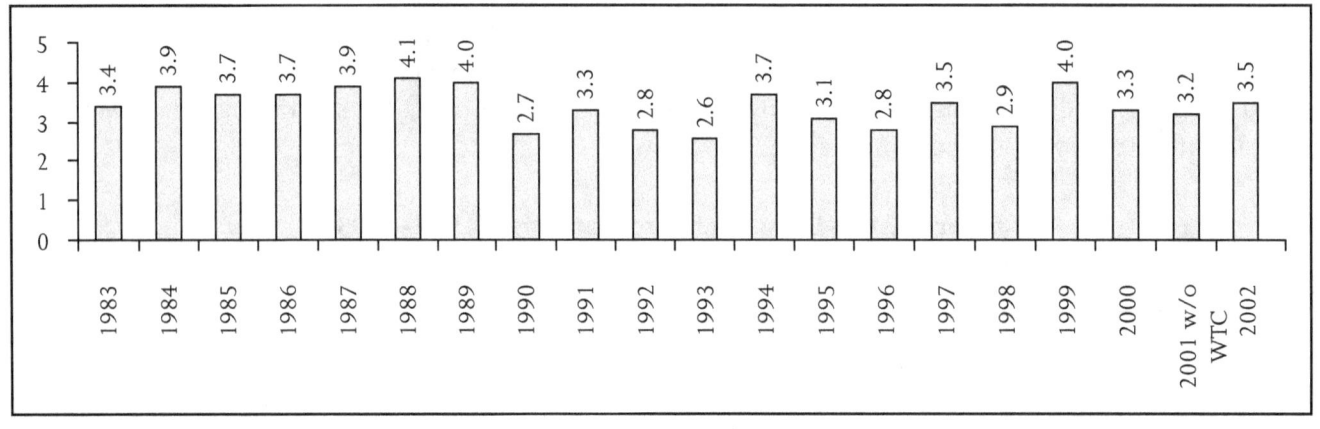

Career and Volunteer Fatalities

Firefighter fatalities in 2003 include 81 volunteer firefighters and 30 career firefighters (Figure 3). Among the volunteer firefighter fatalities, 59 were from local or municipal volunteer fire departments, and 22 were seasonal or contract members of wildland fire agencies. All of the career firefighters who died were members of local or municipal fire departments. Three of the firefighters who died in 2003 were female and 108 were male.

Figure 3. Career versus Volunteer Fatalities

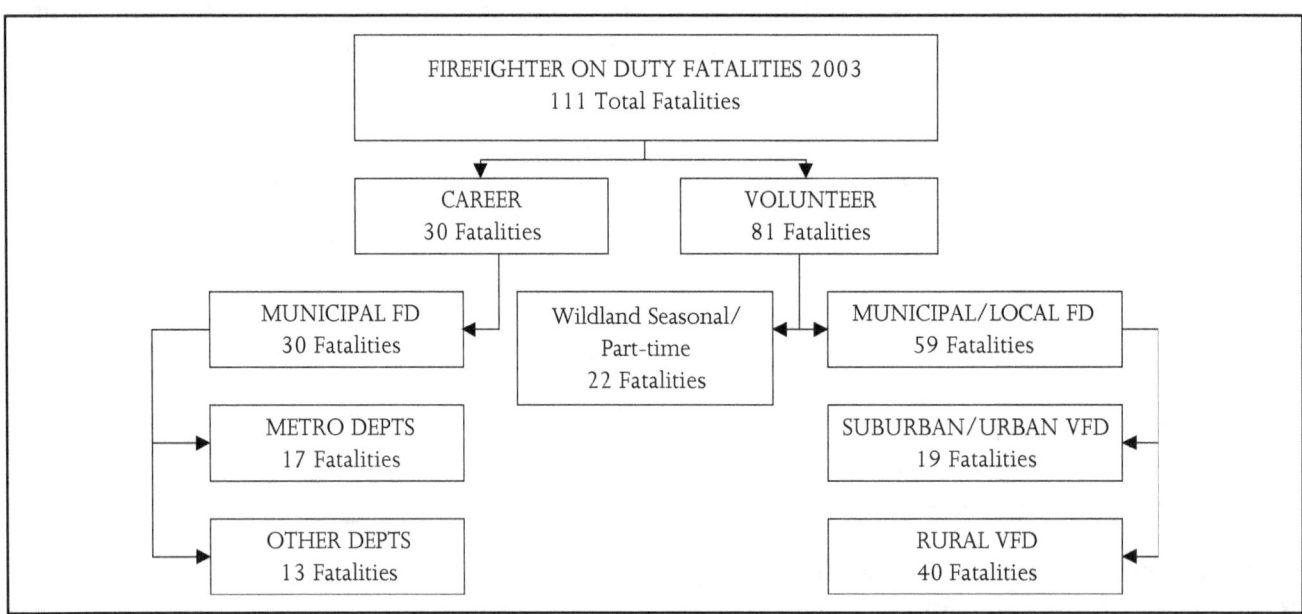

8

Multiple Firefighter Fatality Incidents

The 111 deaths resulted from 98 incidents. There were seven multiple firefighter fatality incidents resulting in the deaths of 20 firefighters.

Table 1. Multiple Firefighter Fatality Incidents

Year	Number of Incidents	Number of Fatalities
2003	7	20
2002	9	25
2001	8	362
2001 w/o WTC	7	18
2000	5	10
1999	6	22
1998	10	22
1997	8	17
1996	3	8

Eight Oregon-based wildland contract firefighters died when their van was involved in a head-on crash with a tractor-trailer truck as they returned to base after fighting wildland fires in Idaho; two Memphis firefighters were killed while they fought an arson-caused fire in a commercial occupancy; two Idaho wildland firefighters were killed when their position was overrun by fire as they constructed a helicopter landing area; two firefighters were killed in Arizona as the result of a helicopter crash at a wildland fire; two Ohio firefighters were killed when a burning silo exploded; two Nevada-based wildland aircraft pilots died in the crash of their airtanker in California; and two Oregon firefighters died when their helicopter become entangled in wires as they conducted a water reconnaissance mission.

In 2003, four fire departments suffered the loss of multiple firefighters in separate incidents. The Charlotte Fire Department lost firefighters as the result of incidents in January and April; the Fire Department City of New York lost firefighters in September and December; the Bureau of Indian Affairs in Whiteriver, Arizona, lost firefighters in May and July; and the Cool Springs Volunteer Fire Department in North Carolina suffered firefighter deaths in July and September.

Wildland Firefighting Fatalities

The number of deaths associated with brush, grass, or wildland firefighting in 2003 was 29. This total includes part-time and seasonal wildland firefighters and municipal or volunteer firefighters engaged in fighting a wildland fire. This is the highest level of firefighter fatalities since 1994 when 14 firefighters perished on Storm King Mountain in Colorado; 36 firefighters died in association with wildland fires that year.

After a year of no burnover deaths in 2002, four firefighters died in 2003 when their positions were overrun by rapid fire progress. Three of these deaths were during fire control operations and one occurred at a planned burn.

Eight firefighters based in Oregon died in the crash of a passenger van; two Idaho firefighters were killed when their position was overrun with fire as they constructed a helicopter landing area; six firefighters died of heart attacks at wildland incidents; four firefighters died in vehicle crashes, two in tankers, one in a pumper-tanker, and one in a pickup truck; and firefighters in Arizona and Colorado were killed when fire advanced rapidly and overcame their ability to escape.

In 2003, there were seven firefighter deaths associated with wildland aircraft firefighting duties. This total includes fixed-wing aircraft and helicopters. Two Arizona firefighters died in a helicopter crash in July, two Nevada-based airtanker pilots died in California in October, two Oregon firefighters died in a helicopter crash in October, and an Oregon-based firefighter died as the result of a helicopter crash in Washington State in July. This is the highest number of firefighter deaths due to aircraft crashes since nine firefighters were killed in 1994.

Table 2. Firefighter Fatalities Associated with Wildland Firefighting

Year	Number of Fatalities
2003	29
2002	23
2001	15
2000	17
1999	26
1998	14
1997	12
1996	9

Table 3. Wildland Firefighting Aircraft Fatalities

Year	Number of Fatalities
2003	7
2002	6
2001	6
2000	6
1999	0
1998	3
1997	5
1996	0
1995	3
1994	9

2003 had the highest level of wildland fire-related firefighter deaths and the highest number of aircraft-related deaths since 1994.

TYPE OF DUTY

Activities related to emergency incidents resulted in the deaths of 78 firefighters (Figure 4). This includes all firefighters who died while responding to an emergency, while at an emergency scene, or while returning from the emergency incident. Nonemergency activities accounted for 33 fatalities. Nonemergency duties include training, administrative activities, or performing other functions that are not related to an emergency incident. A multiyear historical perspective concerning the percentage of firefighter deaths that occurred during emergency duty is presented in Table 4.

Figure 4. Firefighter Fatalities by Type of Duty (2003)

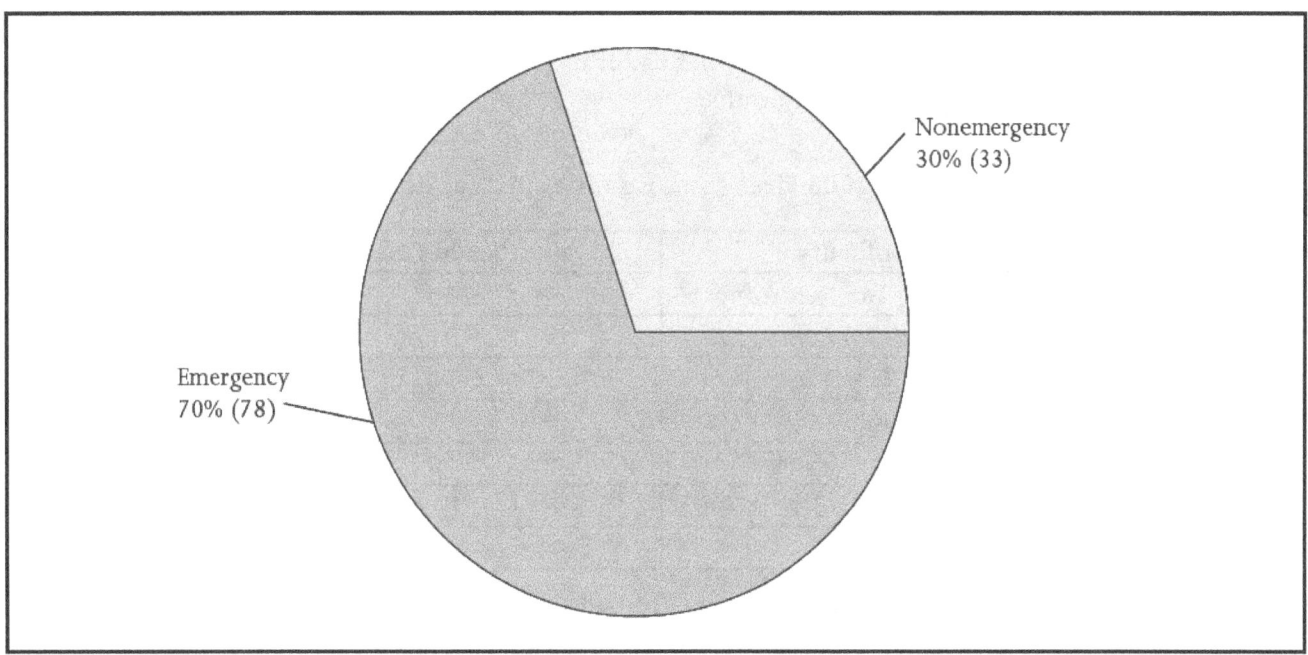

Nonemergency
30% (33)

Emergency
70% (78)

Table 4. Emergency Duty Firefighter Fatalities

Year	Percentage of All Fatalities
2003	70%
2002	73%
2001	65%
2001 w/o WTC	92%
2000	71%
1999	87%
1998	77%
1997	81%
1996	72%

The number of deaths by type of duty being performed in 2003 is shown in Table 5 and presented graphically in Figure 5. For the first time in over a decade, more firefighters died while responding and returning than for any other type of duty in 2003. There were a total of 36 deaths while responding to or returning from an emergency incident in 2003. Vehicle crashes accounted for 24 deaths, including the eight Oregon firefighters who were killed in a crash involving their van, six firefighters killed in personal vehicle crashes while responding, and four aircraft-related deaths.

Table 5. 2003 Firefighter Fatalities by Type of Duty

Type of Duty	Number of Fatalities
Responding and Returning	36
Fireground Operations	31
Other On-Duty	20
Training	12
Nonfire Emergencies	10
After an Incident	2
Total	**111**

Figure 5. Fatalities by Type of Duty (2003)

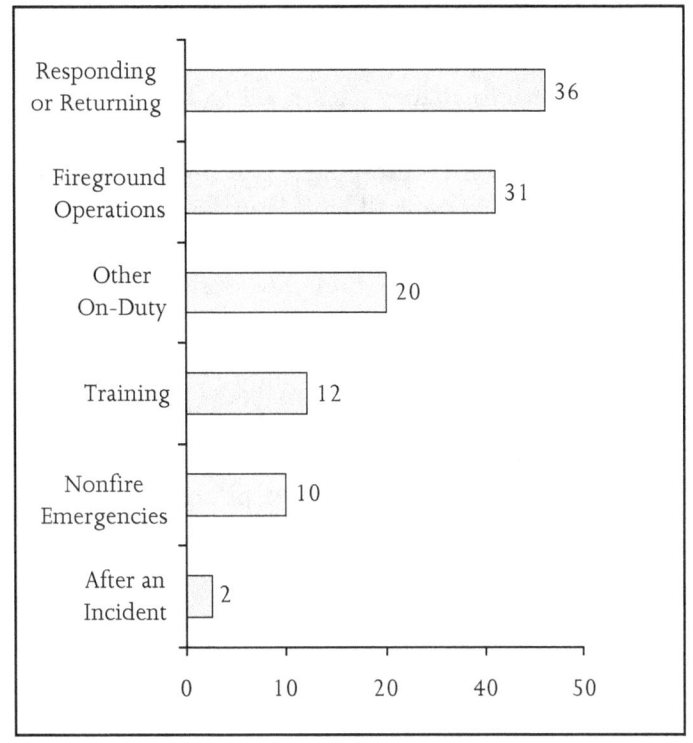

Responding/Returning

Thirty-six firefighters died while responding to or returning from emergency incidents in 2003. Vehicle crashes claimed 24 lives, eight firefighters suffered heart attacks, two firefighters were struck by vehicles, and two firefighters died in falls while responding.

A Wyoming Fire Explorer was killed in the crash of a water tanker. The Fire Explorer was a passenger in the tanker; the driver had a blood alcohol level of .16 at the time of the crash.

The largest number of firefighters killed in a single incident in 2003 occurred when eight Oregon firefighters were killed in a crash as they returned from fighting wildland fires in Idaho. The firefighters were the occupants of a van that crossed the centerline of the roadway to pass a truck. The van was hit head-on by a tractor-trailer truck traveling in the opposite direction. Firefighters in the van had been consuming alcohol but a reliable assessment of the driver's blood alcohol level had not been established by the time this report was prepared.

Six firefighters died in 2003 as they responded to emergencies in their personal vehicles. The deceased firefighter was not wearing a seatbelt in four of the five cases where the status of seatbelt usage was known. Two of the firefighters killed in personally owned vehicle (POV) crashes were under 20 years of age, two were in their 20's, and two were age 30 or older.

Two firefighters were killed in each of two aircraft crashes. In one, a helicopter crashed in Arizona and killed its pilot and a helitack firefighter who was a passenger in the aircraft. The second incident involved the crash of a Nevada-based airtanker as it was repositioned from fires in Arizona to fires in California. The airtanker crashed shortly before reaching its intended destination.

Six firefighters were killed as they responded to emergencies in fire department vehicles. Two deaths involved tanker crashes, and crashes involving a pickup truck, a fire department sedan, an engine-tanker, and an ambulance claimed one life each.

Eight firefighters suffered fatal heart attacks as they responded to or returned from emergencies. Two fire police officers experienced heart attacks as they drove to emergencies, two firefighters had heart attacks immediately upon returning to their stations after a response, one firefighter collapsed just after his arrival at the scene of an emergency, one firefighter became ill while responding and was provided with medical attention when he arrived at the scene, one firefighter was struck with a heart attack after starting the engine on his apparatus prior to leaving the station, and one firefighter suffered a heart attack while hiking in to the scene of a wildland fire.

Two firefighters were killed in falls while responding, including a California career firefighter who fell from a responding engine and a New York volunteer firefighter who fell from a ladder at home after being dispatched to an emergency. One firefighter was struck by a passing vehicle as he crossed the roadway at the scene of a vehicle crash and one firefighter was crushed by a pumper as it backed into quarters after a fire response.

Table 6. Firefighter Fatalities While Responding to or Returning From an Incident

Year	Number of Fatalities
2003	36
2002	13
2001	23
2000	19
1999	26
1998	14
1997	21
1996	22

Fireground Operations

Thirty-one firefighters died while engaging in activities at the scene of a fire in 2003. There were three multiple firefighter fatality incidents that took place on the fireground. Two Memphis firefighters died in an arson-caused fire in a discount store, two Idaho firefighters were killed when rapidly advancing fire overtook their position as they constructed a helicopter landing zone during a wildfire, and two Ohio firefighters were killed when they were thrown to the ground from the top of a burning silo containing scrap wood and sawdust that suddenly exploded.

Lancaster, Massachusetts, firefighter Martin H. McNamara, V, was killed while he fought a fire in the basement of this house.

Sixteen firefighters suffered fatal heart attacks at the fire scene. Seven heart attacks occurred in relation to structure fires, four were related to wildland fires, two were suffered as the result of nonstructural fires, two were fire police officers at fire scenes, and one occurred during the response to an automatic alarm.

In addition to the Memphis firefighters described above, five firefighters were killed while in the interior of burning structures. A firefighter in Pennsylvania was killed after the collapse of a chimney, an Ohio firefighter was burned in a residential structure fire, a New York firefighter was killed in a commercial building fire, a Massachusetts firefighter was killed in a residential structure fire, and a Texas firefighter was killed during a fire in a specialty automotive occupancy.

A California firefighter was overrun by rapid fire progress at a fire near San Diego, and suffered fatal burns; a North Carolina firefighter suffered a knee injury and died later in the year as a complication of knee surgery; a Minnesota firefighter was killed when he was struck by a civilian vehicle at the scene of a fire incident; and an Oregon-based helicopter pilot was killed in a crash while fighting a wildland fire in the State of Washington.

Training

Twelve firefighters died while engaged in training. Three firefighters died when they suffered heart attacks while performing physical fitness duties on duty. A North Carolina firefighter suffered a heart attack shortly after taking command of a live-fire structural fire training exercise, and a North Carolina firefighter suffered a heart attack during an annual wildland firefighting certification test.

Five firefighters were killed in vehicle-related incidents during training. A Maryland firefighter died when his vehicle was struck and pushed off of the road as he returned from a paramedic training class, a Tennessee firefighter fell from the tailgate of a pickup truck as it was driven in a training facility after the completion of training, a Louisiana firefighter died in the crash of a water tanker during training, an Oregon firefighter was killed when the pumper in which he was riding left the roadway and struck a tree, and a Louisiana firefighter was killed when he was struck by a civilian vehicle after dismounting his fire truck (to retrieve a part that had fallen off) as he returned from training.

Two firefighters died during recruit training: one in Texas and one in Florida. In addition to other factors, both firefighters had heart conditions that were difficult to detect in a normal physical examination.

Table 7 offers a multiyear perspective on training fatalities.

Table 7. Firefighter Fatalities During Training

Year	Number of Fatalities
2003	12
2002	11
2001	14
2000	13
1999	3
1998	12
1997	5
1996	6

Nonfire Emergencies

Ten firefighters died at nonfire emergencies. Seven of the deaths were heart attacks, two firefighters were struck by civilian vehicles as they worked on emergency scenes, and one firefighter was killed in the crash of a Texas helicopter during the recovery efforts for the Space Shuttle Columbia.

After the Incident

Two firefighters suffered heart attacks after the conclusion of their work shift. A Massachusetts firefighter working on a wildland fire in Montana died during the night after complaining of feeling unwell the night before and a Maryland firefighter died shortly after going off duty.

Other On-Duty

A total of 20 firefighters died while on duty but their deaths were not associated with the response to any particular emergency. One multiple firefighter fatality incident took the lives of two Oregon firefighters. The firefighters were conducting a reconnaissance mission using a helicopter to search for and document water sources for wildland firefighting. The helicopter became entangled in wires crossing a river and crashed into the water below.

Thirteen firefighters died as the result of heart attacks suffered on duty. Five career firefighters suffered heart attacks while on duty, two volunteer firefighters were struck with heart attacks as they attended to fire department business in the fire station, two firefighters suffered heart attacks while working on fire department fundraising events, two fire police officers died while directing traffic for parades, one firefighter suffered a heart attack while clearing brush behind the fire station, and one firefighter became ill after responding to a crash involving a fire truck.

An Arizona firefighter was killed when a fire progressed unexpectedly at a controlled burn, a Nebraska firefighter was killed while transporting a patient by ambulance when the vehicle was struck from behind by a tractor-trailer truck, an Idaho firefighter was killed in a vehicle crash as he gathered supplies for an extended wildland fire fight, an Illinois firefighter was killed as the result of a crash involving the lawn tractor that he was operating, and a Pennsylvania firefighter died during a standby at his fire station when he fell from a ladder.

Career, Volunteer, and Wildland Fatalities by Type of Duty

Figure 6 depicts career, volunteer, and wildland firefighter deaths by type of duty. Wildland career, wildland seasonal, and wildland contractor deaths were grouped together. As in past years, there were a disproportionate number of fatalities experienced by volunteer firefighters responding to and returning from alarms as compared to career firefighters. The large number of wildland firefighters in the *responding to and returning from* category reflects the deaths of eight Oregon firefighters in a single incident.

Table 8. Career, Volunteer, and Wildland Fatalities by Type of Duty

	Fireground Operations	Nonfire Emergency	Responding/ Returning from Alarm	Training	Other On-Duty	After Incident
Career	12	0	4	6	7	1
Volunteer	16	9	19	5	10	0
Wildland	3	1	13	1	3	1

Figure 6. Career, Volunteer, and Wildland Fatalities by Type of Duty (2003)

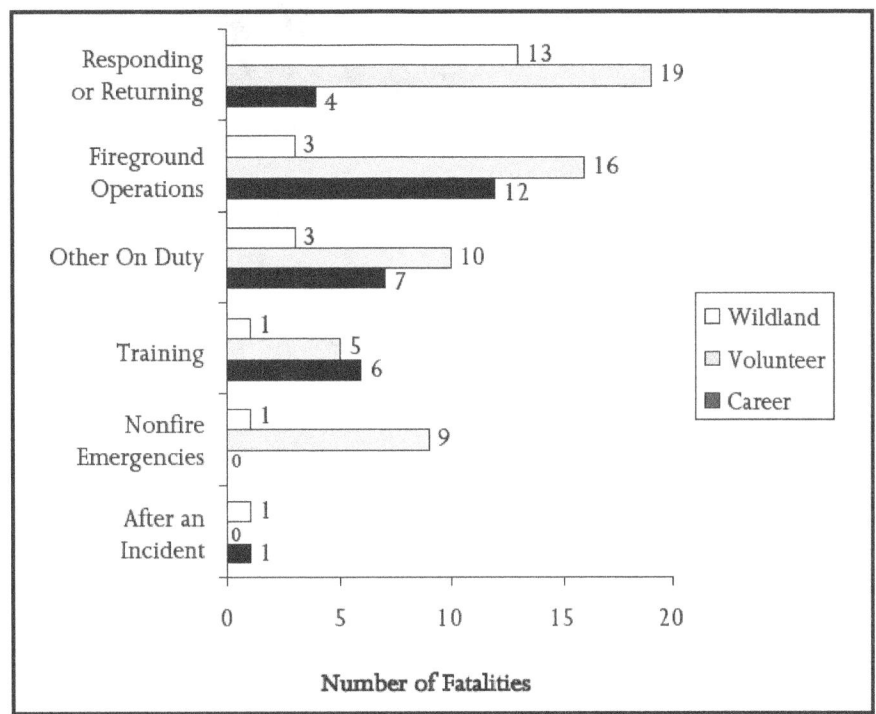

Type of Emergency Duty

In 2003, 65 firefighters died while engaged directly in the delivery of emergency services. This number includes deaths that were the result of injuries sustained on the incident scene or en route to the incident scene, and firefighters who became ill on an incident scene and later died. It does not include firefighters who became ill or died while returning from an incident (such as a vehicle collision while returning from an incident). Figure 7 shows the number of firefighters killed in firefighting, emergency medical services, technical-rescue-related incidents, and other emergency incidents in 2003.

Forty-six firefighters were killed in relation to fires, 16 firefighters were killed in relation to EMS calls, two firefighters were killed at emergencies that involved hazardous materials, and one firefighter died while responding to a technical rescue (water rescue).

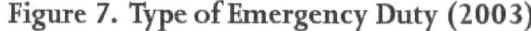

Figure 7. Type of Emergency Duty (2003)

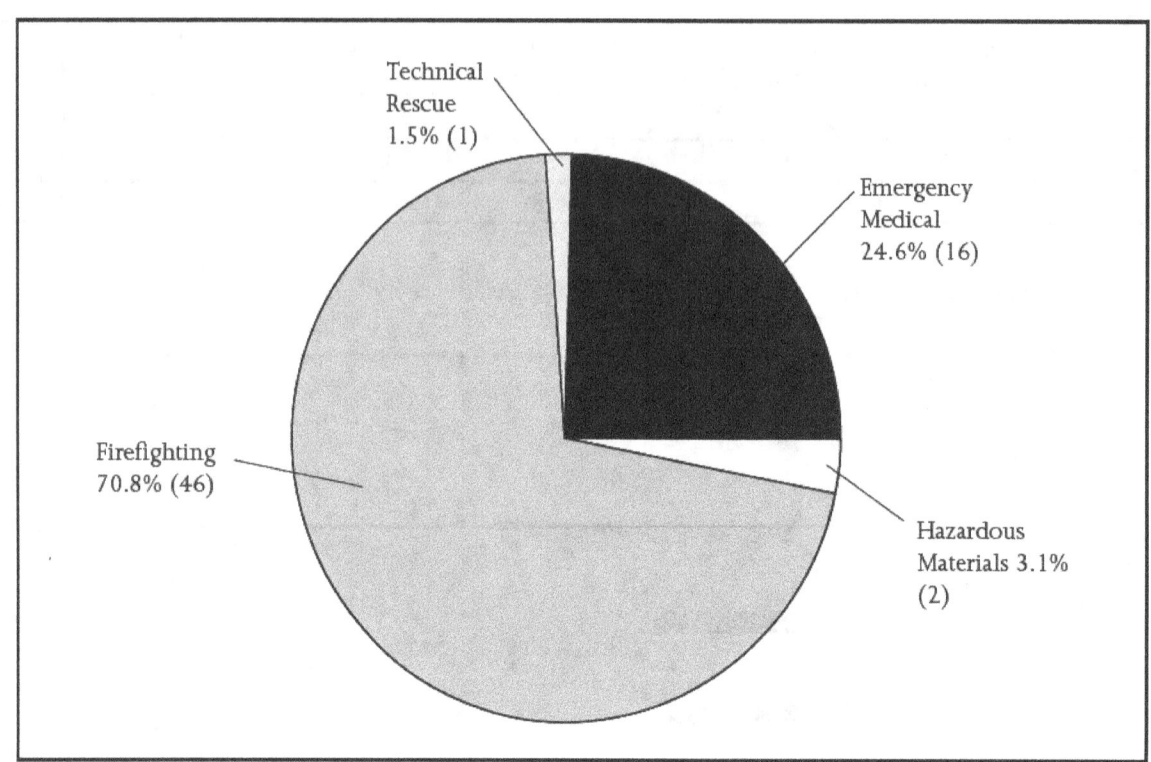

Technical
Rescue
1.5% (1)

Emergency
Medical
24.6% (16)

Firefighting
70.8% (46)

Hazardous
Materials 3.1%
(2)

CAUSE OF FATAL INJURY

The term *cause of injury* refers to the action, lack of action, or circumstances that resulted directly in the fatal injury. The term *nature of injury* refers to the medical cause of the fatal injury or illness, often referred to as the physiological cause of death. A fatal injury usually is the result of a chain of events; the first of which is recorded as the cause.

Table 9 and Figure 8 show the distribution of deaths by cause of fatal injury or illness.

Table 9. Cause of Fatal Injury (2003)

Cause	Number of Fatalities
Stress/Overexertion	51
Vehicle Collision	34
Caught/Trapped	11
Fall	8
Struck by	5
Other	2
Total	**111**

Figure 8. Fatalities by Cause of Fatal Injury (2003)

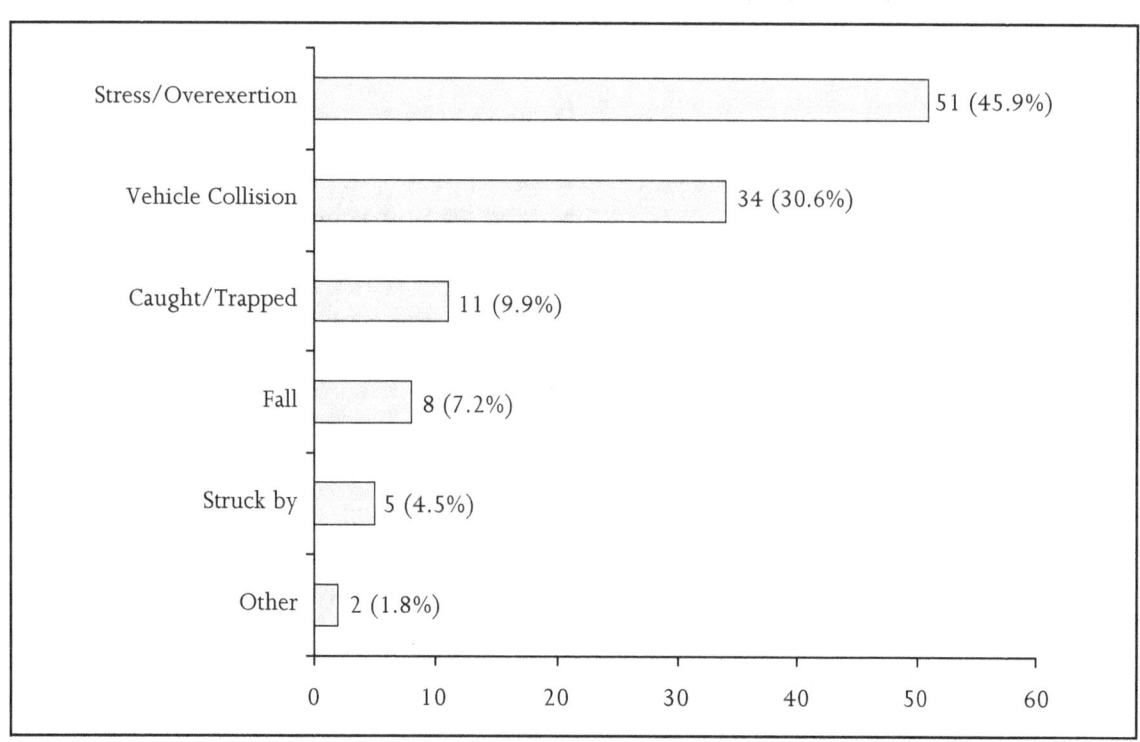

Stress or Overexertion

Firefighting is extremely strenuous physical work and is likely one of the most physically demanding activities that the human body performs. As it has been in every year for more than the past decade, the largest cause of firefighter deaths in 2003 was stress or overexertion. Stress or overexertion was listed as the primary factor in 51 firefighter deaths in 2003.

Most firefighter deaths attributed to stress result from heart attacks. Of the 51 stress-related fatalities in 2003, 50 firefighters died of heart attacks and one died of heat exhaustion.

Stress or overexertion-related firefighter deaths in 2003 included 20 firefighters who were on duty, but not working on an emergency incident, 10 firefighters working in association with structure fires, seven firefighters who were working in relation to motor vehicle crashes, six in association with wildland fires, four in response to outside (nonstructural) fires, two in association with automatic fire alarms, one at an EMS incident, and one in relation to a hazardous materials incident.

Table 10. Fatalities Caused by Stress or Overexertion

Year	Number	Percent of All Fatalities
2003	51	45.9%
2002	38	38.0%
2001	43	40.9%*
2000	46	44.6%
1999	56	49.5%
1998	43	46.2%
1997	41	41.0%
1996	46	46.4%

* Does not include the firefighter deaths of September 11, 2001, in New York City.

Vehicle Collisions

The second leading cause of fatal injury for firefighters who died in 2003 was vehicle collisions. This cause is usually the second most common cause of firefighter fatalities. A total of 34 firefighters died in 2003 in vehicle crashes. The number of firefighters killed in vehicle collisions has been steadily rising (see Figure 9). The 34 firefighters killed in vehicle collisions in 2003 is the highest single-year total since 1990.

There were four multiple-firefighter fatality vehicle collisions in 2003. Eight Oregon-based firefighters were killed in a vehicle collision as they returned to Oregon after fighting wildland fires out of State. This is the second consecutive year with a multiple-firefighter fatality incident involving wildland firefighters traveling in vans. Two Oregon firefighters died when their helicopter became entangled in wires strung across a river as they scouted water supply locations, two firefighters died in California in the crash of an airtanker, and two Arizona firefighters were killed when their helicopter crashed as they worked a wildland fire.

In 2003, 6 firefighters died in Personally Owned Vehicle (POV) crashes while responding to emergencies and a Maryland firefighter died when his POV was struck by another vehicle and left the highway as he

> From 1990 through 2003, 50 firefighters died in POV crashes — an average of just under four deaths per year.

returned from training. All of the 6 firefighters killed while responding were volunteers, 5 were under 30 years of age, and only 1 was found to be wearing a seat belt at the time of the crash.

A Kansas firefighter died in the crash of a fire department sedan while responding to an incident; an Iowa firefighter died when his city-owned pickup truck left the roadway and crashed as he brought supplies to a wildland fire; and a Louisiana firefighter died when his fire department pickup was struck by a train as he responded to an EMS incident.

Figure 9. Firefighter Fatalities in Vehicle Collisions

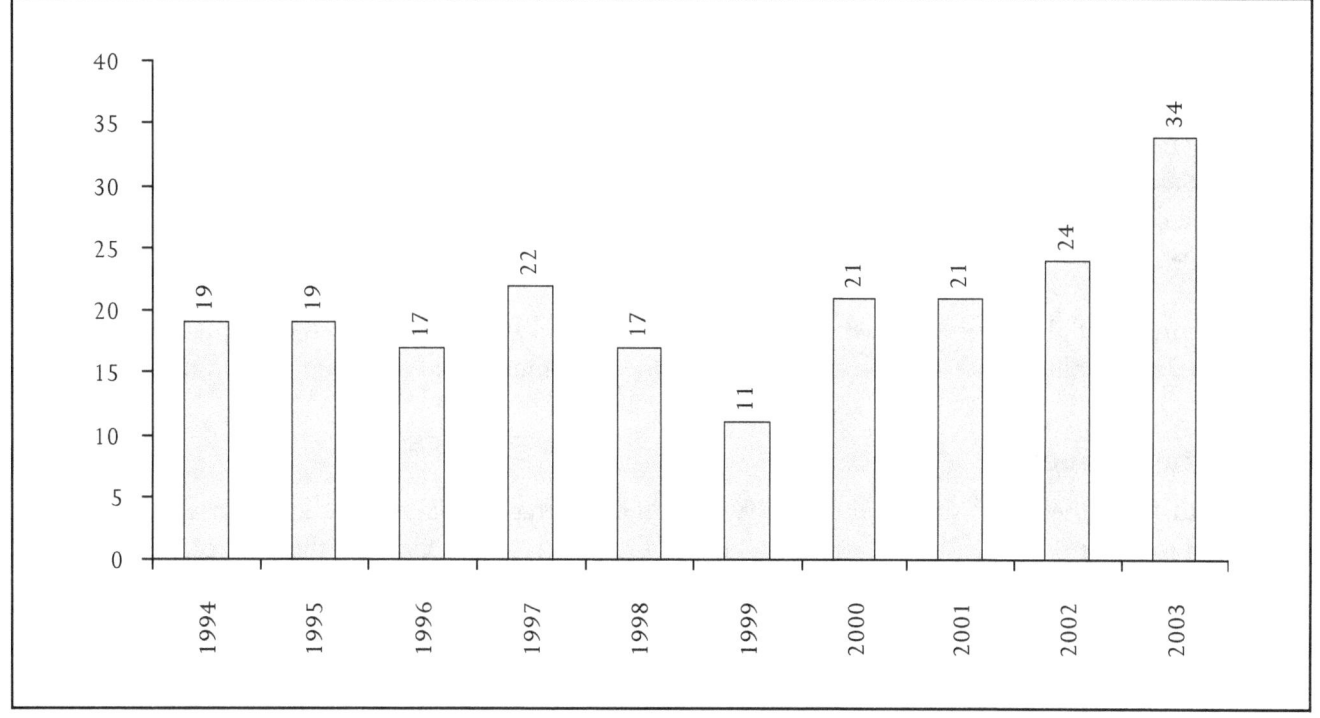

Two firefighters were killed in crashes involving ambulances. A Texas firefighter who was a front-seat passenger in an ambulance was killed when the ambulance was involved in a head-on collision with a car that had crossed the centerline of the roadway, and a Nebraska firefighter who was driving an ambulance was killed when the ambulance was struck from behind by a tractor-trailer truck as the ambulance slowed to pass through a construction zone.

In addition to the six aircraft-related deaths described above, two firefighters died in aircraft crashes in 2003. A Texas firefighter was killed in a helicopter crash as he was engaged in the Space Shuttle Columbia recovery mission, and an Oregon firefighter was killed in a helicopter crash while fighting a wildland fire in Washington State.

Five firefighters were killed in fire apparatus crashes. A Louisiana firefighter was killed in the crash of a tanker during driver training; a number of mechanical problems contributed to the crash. An Oregon firefighter riding in the right-front-seat of a pumper was killed when the apparatus left the roadway and crashed into a tree. A New Mexico firefighter was killed as he drove a military surplus truck that had been converted to a tanker to a wildland fire; faulty brakes contributed to the crash. A firefighter in Ohio was killed when the engine-tanker that he was driving rolled, and a Wyoming firefighter drowned when the tanker in which she was a passenger rolled off the road.

Scene of the crash that claimed the life of Firefighter Richard A. Long of the Gallipolis (Ohio) Volunteer Fire Department. The apparatus left the right side of the roadway, came back onto the road, and rolled.

A West Virginia firefighter was killed when he was crushed by a pumper as it backed into quarters. Two firefighters had been rolling hose in front of the station; one was injured and the other died in the incident.

Caught or Trapped

In 2003, 11 firefighters were killed when they were caught or trapped. Seven firefighters died in structural fires when they were trapped by fire progress. Rapid fire progress also claimed the lives of four wildland firefighters.

Two Memphis firefighters were killed when fire developed rapidly and cut off escape routes during a structure fire in a commercial occupancy, a Texas firefighter was killed while fighting a fire with explosions in a specialized vehicle restoration shop, a Cincinnati firefighter was killed when fire progressed rapidly in a single-family residential structure fire, a Pennsylvania firefighter was trapped by falling structural members in a structural fire, a Fire Department City of New York firefighter was killed when he became separated from his crew and trapped in a commercial building fire, and a Massachusetts firefighter was killed while fighting a fire in the basement of a multifamily residence.

Two Idaho firefighters were killed when fire overran their position as they constructed a helicopter landing zone, an Arizona firefighter was severely burned when he was overrun by rapidly advancing fire at a controlled burn, and a California firefighter was killed during a major wildland fire when he and his crew were trapped by a rapidly advancing fire.

Falls

Eight firefighters died in 2003 as the result of falls. Two Ohio firefighters were killed in the explosion of a silo that was involved with fire; one firefighter had been on top of the silo and the other was in the bucket of an aerial tower ladder when the explosion occurred. A California firefighter died after she fell from the jump seat of a pumper as it responded to an emergency, and a New York firefighter died when he fell from a ladder at home as he began his response to an emergency.

Two firefighters died in falls at the fire station in 2003. A Pennsylvania firefighter died during a standby at the fire station when he fell from a stepladder, and an Illinois firefighter died when the garden tractor he was driving crashed into a ditch outside of the fire station.

Finally, a New Jersey firefighter died of head injuries he suffered when he fell at the scene of a motor vehicle crash, and a Tennessee firefighter died when he fell from the tailgate of a moving pickup truck on the grounds of a training facility.

Struck by Object

Being struck by an object was the fifth leading cause of fatal firefighter injuries in 2003. Five firefighters died after being struck by vehicles while on duty.

Four firefighters were struck at the emergency scene. A Louisiana firefighter was killed while standing by at a crash involving hazardous materials; the weather was foggy and a vehicle swerved to avoid traffic barriers and struck the firefighter. A Texas firefighter was killed after stopping at the scene of a crash on a divided highway; the firefighter was struck by a tractor-trailer truck after crossing the median to reach the crash

St. Cloud, Minnesota, Assistant Chief Don Billig was killed when he was struck by a pickup truck at the scene of an emergency.

scene. A Minnesota firefighter was killed as he and other firefighters replaced construction zone barriers after a response; a pickup truck operated by a driver who likely was impaired crashed into the firefighters and their vehicle. A New Jersey fire police officer was killed when he was struck by a vehicle as he performed traffic control duties at the scene of an earlier crash.

A Louisiana firefighter was killed when he was struck by a passing vehicle. Firefighters were returning from training when they stopped to retrieve a piece that had fallen from one of their vehicles.

Other

Two firefighters died in 2003 of causes that are not categorized above. A Guam firefighter died after responding to several emergencies; the cause of death was pulmonary edema following smoke exposure. A Charlotte, North Carolina, firefighter died of a pulmonary embolism that resulted from knee surgery to repair an on-duty knee injury.

NATURE OF FATAL INJURY

Table 11 and Figure 10 show the distribution of the 111 deaths by the medical nature of the fatal injury or illness.

Table 11. Nature of Fatal Injury (2003)

Nature	Number of Fatalities
Heart Attack	50
Internal Trauma	41
Burns	8
Asphyxiation	6
Crushed	2
Drowning	1
Heat Exhaustion	1
Other	2
Total	**111**

Figure 10. Fatalities by Nature of Fatal Injury (2003)

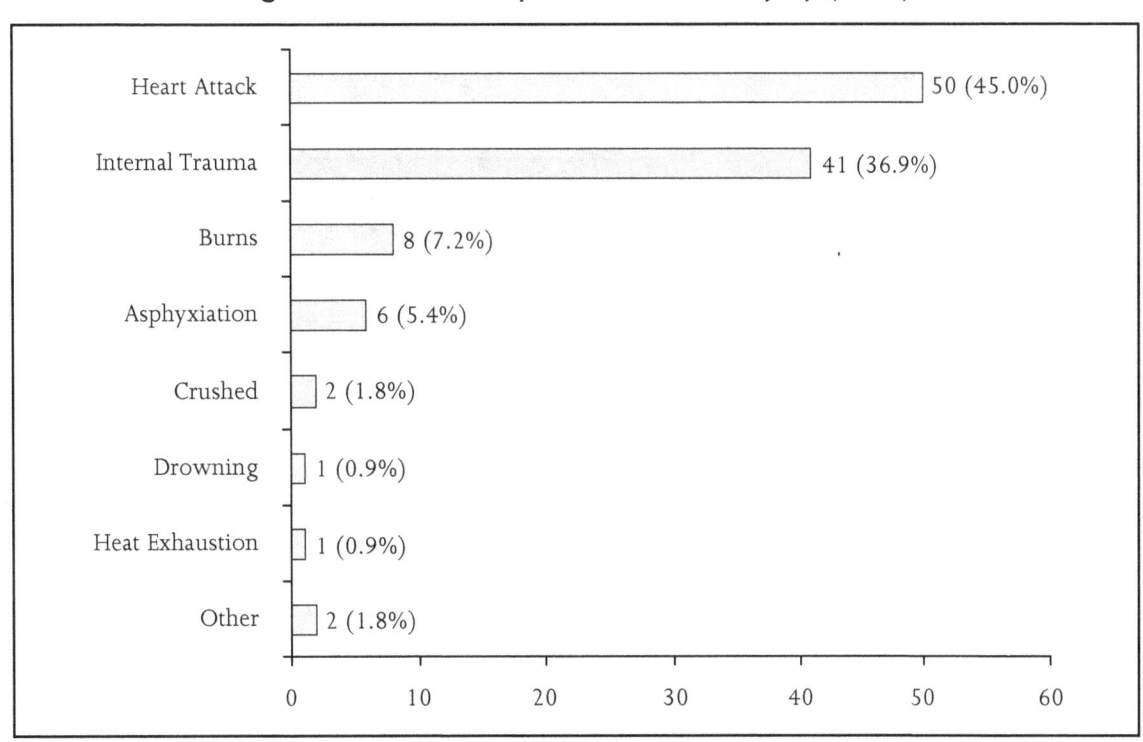

Heart Attack

In 2003, heart attacks were the nature of death for 50 firefighters. Heart attacks are usually the leading nature of firefighter deaths. Figure 11 provides a detailed breakdown of heart attacks by type of duty.

Figure 11. Heart Attack by Type of Duty (2003)

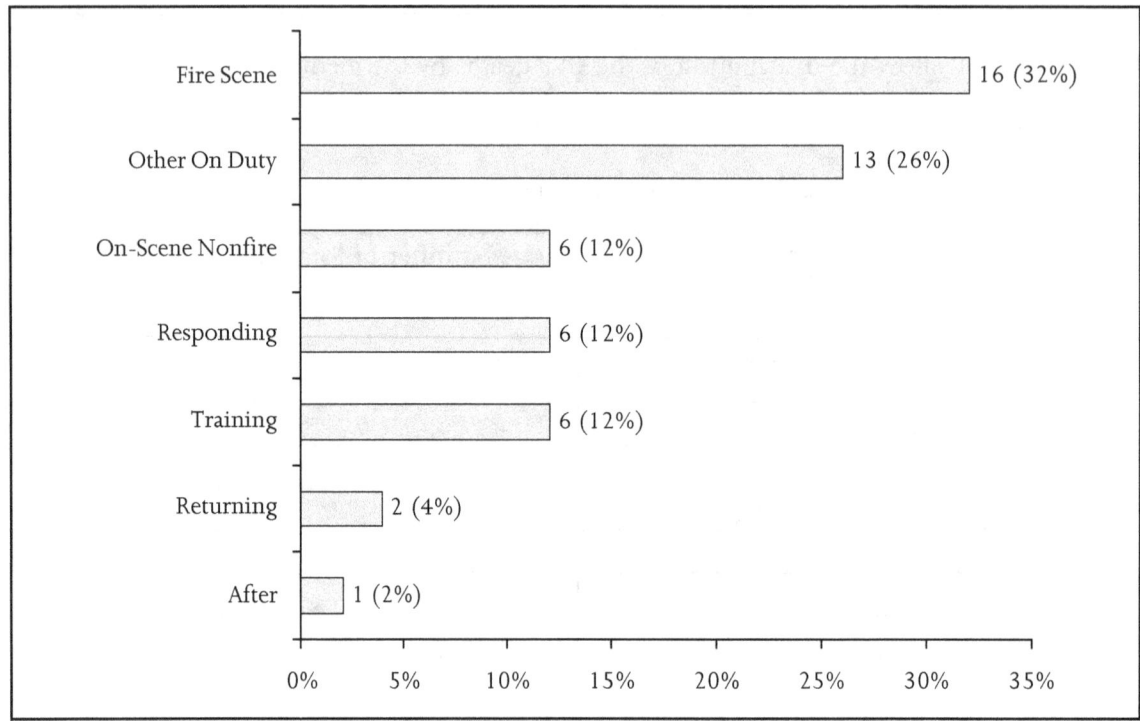

Sixteen firefighters died of heart attacks suffered while on the scene of a fire. Seven firefighters suffered heart attacks while working on the scene of structure fires, four firefighters were engaged in wildland firefighting, two firefighters died of heart attacks suffered on the scene of nonstructural fires, a fire police officer experienced a heart attack on the scene of a structure fire, a fire police officer collapsed on the scene of a vehicle fire, and a firefighter became ill while responding to the scene of an automatic fire alarm activation.

> The median age of the firefighters who died of an on-duty heart attack in 2003 was 52. The youngest victim was 35 years old and the oldest was 81 years of age.

Thirteen firefighters were struck with heart attacks while on duty but not while completing any incident or training-related duties. Six firefighters suffered heart attacks during normal station duties, two firefighters became ill while working on fire department fundraising events, two fire police officers died while working parade events, one firefighter suffered a heart attack while clearing brush near the fire station, one firefighter died of a heart attack at home after a busy shift of fire and EMS responses, and a Texas firefighter suffered a heart attack after working at the scene of a nonfatal fire apparatus crash.

Six firefighters died of heart attacks that occurred at nonfire emergencies. Four of the six deaths occurred at the scene of motor vehicle crashes. An Idaho firefighter suffered a heart attack at the scene of a carbon monoxide incident at a residence. Finally, a Wisconsin firefighter became ill at the scene of a medical emergency.

Six firefighters died of heart attacks that occurred while they were responding to emergencies. Two fire police officers experienced heart attacks while responding to emergencies; their vehicles crashed subsequent to their attacks. One firefighter became ill after starting the engine on his apparatus, one firefighter became ill while responding in a fire department vehicle and collapsed of a heart attack on the scene, and one firefighter dismounted his apparatus at a motor vehicle crash and immediately collapsed. An Idaho firefighter suffered a heart attack as he and other firefighters hiked into the location of a wildland fire.

Heart attacks struck six firefighters during training. Three firefighters became ill while performing on-duty fitness activities. One firefighter died during a recertification pack test for wildland firefighters, a Dallas firefighter died while engaged in recruit training, and one firefighter died of a heart attack on the scene of a live-fire structural training fire.

Two firefighters died while returning from incidents. Both firefighters returned to their stations, stowed their gear, and collapsed. One Massachusetts firefighter died after the conclusion of his workday; the firefighter was found in his tent by other firefighters.

Internal Trauma

In 2003, internal trauma was the nature of death responsible for 41 firefighter deaths. With the exception of 2001 and the traumatic deaths of hundreds of firefighters on a single day, this is the highest number of deaths of this nature since 1990.

Table 12. Internal Trauma Firefighter Fatalities

Year	Number of Fatalities
2003	41
2002	34
2001	28*
2000	36
1999	25
1998	27
1997	32
1996	32

*Does not include the firefighter fatalities of September 11, 2001, in New York City.

Eight traumatic deaths were the result of aircraft crashes, including three multiple firefighter fatality incidents. Six firefighters died in POV crashes while responding, and one firefighter died in a POV crash while returning from training. Six of the eight Oregon wildland firefighters died of traumatic injuries in a van crash.

Five firefighters were struck by vehicles while working on emergency scenes and fire department vehicle crashes claimed another eight firefighters.

Seven firefighters died in falls. Two falls were from vehicles, two falls were from ladders, two firefighters fell from a height as the result of a silo explosion, and one firefighter fell at the scene of a vehicle crash and struck his head.

Burns

Eight firefighters died as the result of burns in 2003: four in wildland fires and four in structure fires. Two Idaho firefighters suffered fatal burns when their position was overrun by rapidly advancing fire, an Arizona firefighter received fatal burns when a controlled burn flared up and overran his position, and a California firefighter was overrun and burned during a major wildland fire in the San Diego area.

Two Memphis firefighters died as the result of burns received in a commercial building structure fire, a Cincinnati firefighter suffered burns when he was trapped in a flashover, and a Texas firefighter died of burns received during a fire in a specialty auto supply store.

Asphyxiation

Asphyxiation was the fourth leading medical reason for firefighter deaths in 2003, responsible for six deaths. Two firefighters died of smoke and soot inhalation in fires in Massachusetts and New York. A Kansas firefighter died of positional asphyxiation after a vehicle crash, and a Pennsylvania firefighter died of compressional asphyxiation after a building collapse. Two of the eight Oregon wildland firefighters killed in a van crash died of asphyxiation.

Table 13. Firefighter Fatalities Due to Asphyxiation

Year	Number of Fatalities
2003	6
2002	15
2001	18
2000	13
1999	16
1998	15
1997	15
1996	5

Crushed

In 2003, two firefighters died when they were crushed. An Ohio firefighter was crushed when the apparatus he was driving left the roadway and rolled, and a West Virginia firefighter was killed when he was run over by a fire department pumper backing into the fire station after an emergency incident.

Drowning

A Wyoming firefighter drowned after the tanker in which she was a passenger experienced a crash. The apparatus ended up on its side in a ditch; the firefighter was pinned between the apparatus and the ground. Water leaking from the water tank on the apparatus filled the ditch and drowned the firefighter.

Heat Exhaustion

A Florida firefighter died of heat exhaustion during a training exercise. The training scenario involved the use of a training prop designed to resemble a ship. The firefighter also suffered burns in the incident.

Other

Two firefighters died in situations where the nature of their deaths does not fall into any of the categories previously described. One firefighter died of pulmonary edema after fighting a fire and the other firefighter died of a pulmonary embolism resulting from a work-related knee surgery.

FIREFIGHTER AGES

Figure 12 shows the percentage distribution of firefighter deaths by age and nature of the fatal injury. Table 14 provides counts of firefighter fatalities by age and the nature of the fatal injury.

Figure 12. Fatalities by Age and Nature (2003)

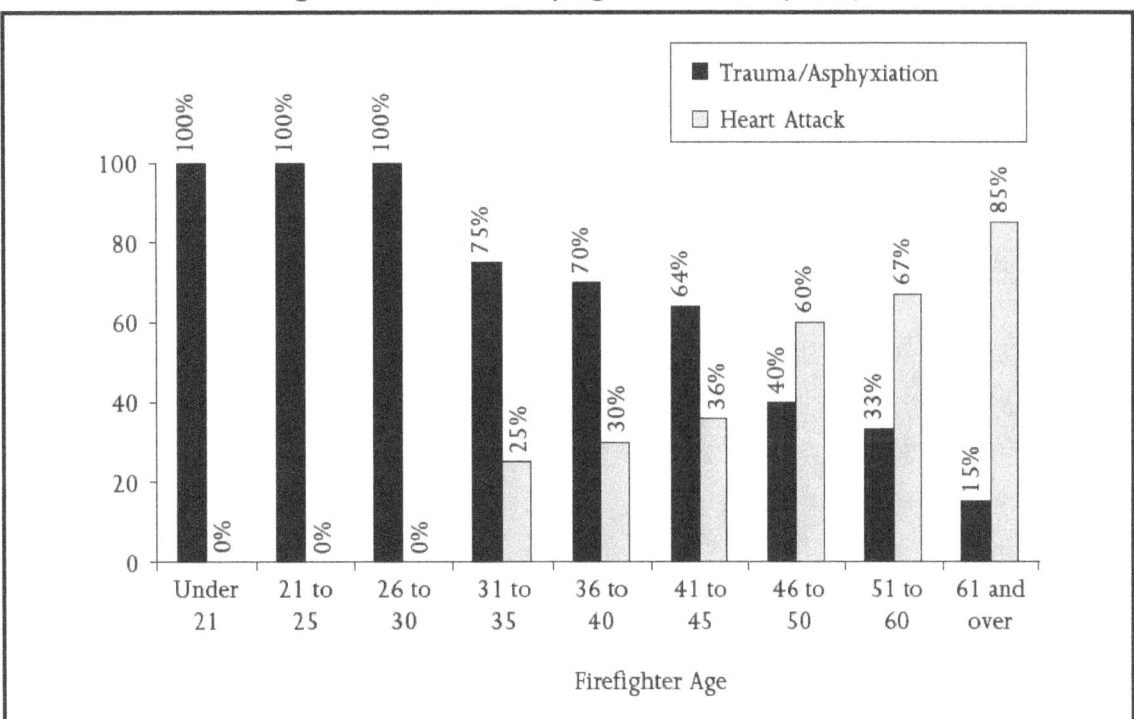

As in most years, younger firefighters were more likely to have died as a result of traumatic injuries such as injuries from an apparatus accident or after becoming caught or trapped during firefighting operations. Stress plays an increasing role in firefighter deaths as age increases.

> The youngest firefighters killed in 2003 were Karlton Briscoe of Mississippi and Anndee Huber of Wyoming, both 16. The oldest firefighter killed was Fire Police Officer Harding Wentzell of Maine, age 81.

Table 14. Firefighter Ages and Nature of Fatal Injury

Age Range	Nontrauma Total	Trauma Total
under 21	0	6
21 to 25	0	11
26 to 30	0	6
31 to 35	1	3
36 to 40	3	7
41 to 45	4	7
46 to 50	12	8
51 to 60	20	10
61 & over	11	2

FIXED PROPERTY USE FOR STRUCTURAL FIREFIGHTING FATALITIES

There were 19 firefighter fatalities in 2003 where the firefighters became ill while on the scene or engaged in structural firefighting. Table 15 shows the distribution of these deaths by fixed property use. As in most years, residential occupancies accounted for the highest number of these fireground fatalities, with 10 deaths. In 2003, however, the number of deaths in commercial structures is very close to the number of firefighter fatalities associated with residential occupancies.

Table 16 shows the number of firefighter deaths in residential occupancies for the past 7 years. Residential occupancies usually account for 70 to 80 percent of all structure fires and a similar percentage of the civilian fire deaths each year*. Historically, the frequency of firefighter deaths in relation to the number of fires is much higher for nonresidential structures.

Table 15. Structural Firefighting Fatalities by Fixed Property Use in 2003

Fixed Property Use	Number	Percent
Residential	10	52%
Commercial	9	48%

Table 16. Firefighter Fatalities In Residential Occupancies

Year	Number of Fatalities
2003	10
2002	21
2001	17
2000	21
1999	23
1998	17
1997	16
1996	19

*Complete 2003 NFIRS fire incidence data were not available at the time of this report, but residential fires typically account for between 70 and 80 percent of all civilian fatalities each year according to the NFPA.

TYPE OF ACTIVITY

Table 17 and Figure 13 show the types of fireground activities firefighters were engaged in at the time they sustained their fatal injuries or illnesses. This total includes all firefighting duties such as wildland firefighting and structural firefighting. In 2003 there were a total of 31 firefighter deaths on the fireground.

Table 17. Type of Activity (2003)

Nature	Number
Fire Attack	11
Suppression Support	6
Water Supply	5
Scene Safety	2
Ventilation	2
Other	5
Total	**31**

Figure 13. Fatalities by Type of Activity (2003)

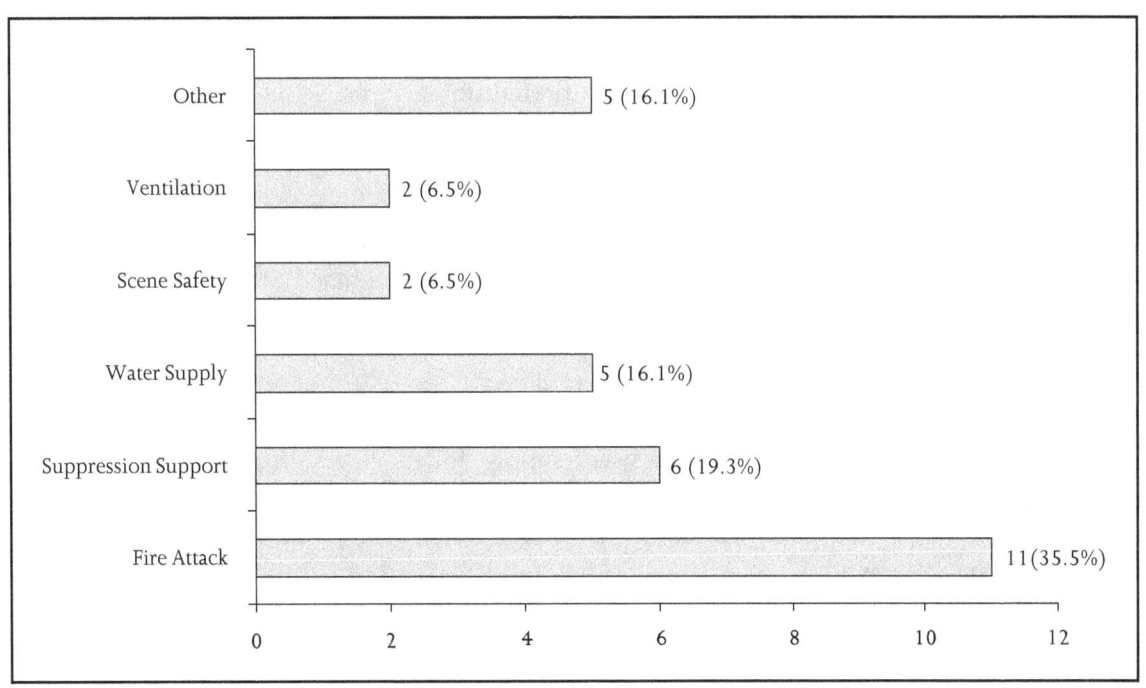

Fire Attack

In 2003, 11 firefighters were killed as they engaged in direct fire attack, such as advancing or operating a hoseline at a fire scene. In years past, most fireground firefighter deaths occur while the firefighter was engaged in fire attack (see Table 18).

Two multiple-firefighter fatality incidents claimed the lives of a total of four firefighters engaged in fire attack in Ohio and Tennessee. Two Ohio firefighters fell from the top of a silo that was involved with fire when an explosion occurred, and two Memphis firefighters died in a commercial structure fire when rapid fire progress and structural collapse cut off their escape routes.

Rapid fire progress claimed the lives of firefighters in California, Ohio, Massachusetts, and Texas. Firefighters in Illinois, Indiana, and New York suffered heart attacks while engaged directly in firefighting duties.

Table 18. Firefighter Fatalities While Engaged in Fire Attack

Year	Number of Fatalities
2003	11
2002	13
2001	13
2000	13
1999	16
1998	18
1997	21
1996	9

Suppression Support

Six firefighters were killed in 2003 as they supported firefighting efforts. Two Idaho firefighters were killed when fire progressed rapidly and overran the helicopter landing zone that they were constructing. A New York firefighter suffered a heart attack after working to overhaul an arson-caused exterior fire and check for extension. A Pennsylvania firefighter was killed while overhauling the interior of a large residence when a structural collapse occurred. A Texas firefighter died after suffering a heart attack as he plowed fire breaks around a wildland fire, and a New Jersey firefighter died as he coordinated efforts at a multi-alarm commercial structure fire.

Water Supply

Five firefighters died in 2003 while engaged in water supply duties. Four of the five deaths involved heart attacks for firefighters assigned as apparatus operators at incident scenes. Two of the heart attacks occurred at structural fires and two occurred at wildland fires. A Charlotte, North Carolina, fire apparatus engineer injured his knee as he stretched hose at a structural fire and later died of complications of surgery to repair the knee injury.

Scene Safety

Two fire police officers died in 2003 at fire scenes. Both officers suffered heart attacks as they performed their duties.

Ventilation

Two firefighters died in 2003 while engaged in ventilation duties on the roof of a burning structure; the fires occurred in Georgia and Pennsylvania. Both fires were human caused: one by an arsonist and the second by a child playing with matches.

Other Duties

Two firefighters were killed performing activities that are not classified. A Minnesota firefighter was struck by a vehicle as he replaced barriers at a construction site after a response into the site for a reported fire and a Connecticut firefighter suffered a heart attack at the scene of an automatic fire alarm.

A Wisconsin fire chief died while commanding a wildland fire. A New York firefighter died while performing search-and-rescue duties during a structural fire in a commercial building, and a Nevada-based firefighter died while operating a helicopter at a wildland fire in the State of Washington.

TIME OF INJURY

The distribution of all 2003 firefighter deaths according to the time of day when the fatal injury occurred is illustrated in Figure 14. The time of death for four firefighters either was not known or was not reported.

Figure 14. Fatalities by Time of Fatal Injury (2003)

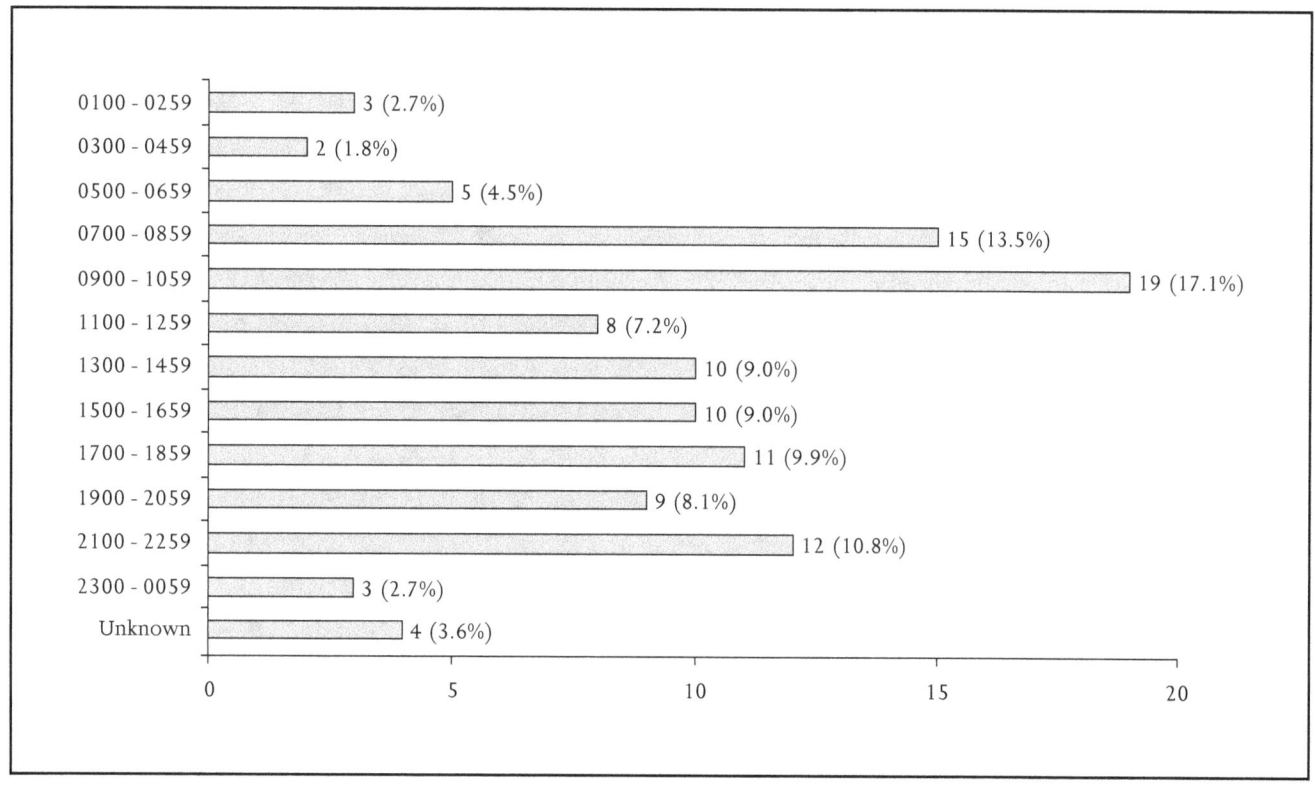

MONTH OF THE YEAR

Figure 15 illustrates firefighter fatalities by month of the year. Firefighter fatalities were highest in August due to a number of wildland firefighting deaths.

Figure 15. Fatalities by Month of the Year (2003)

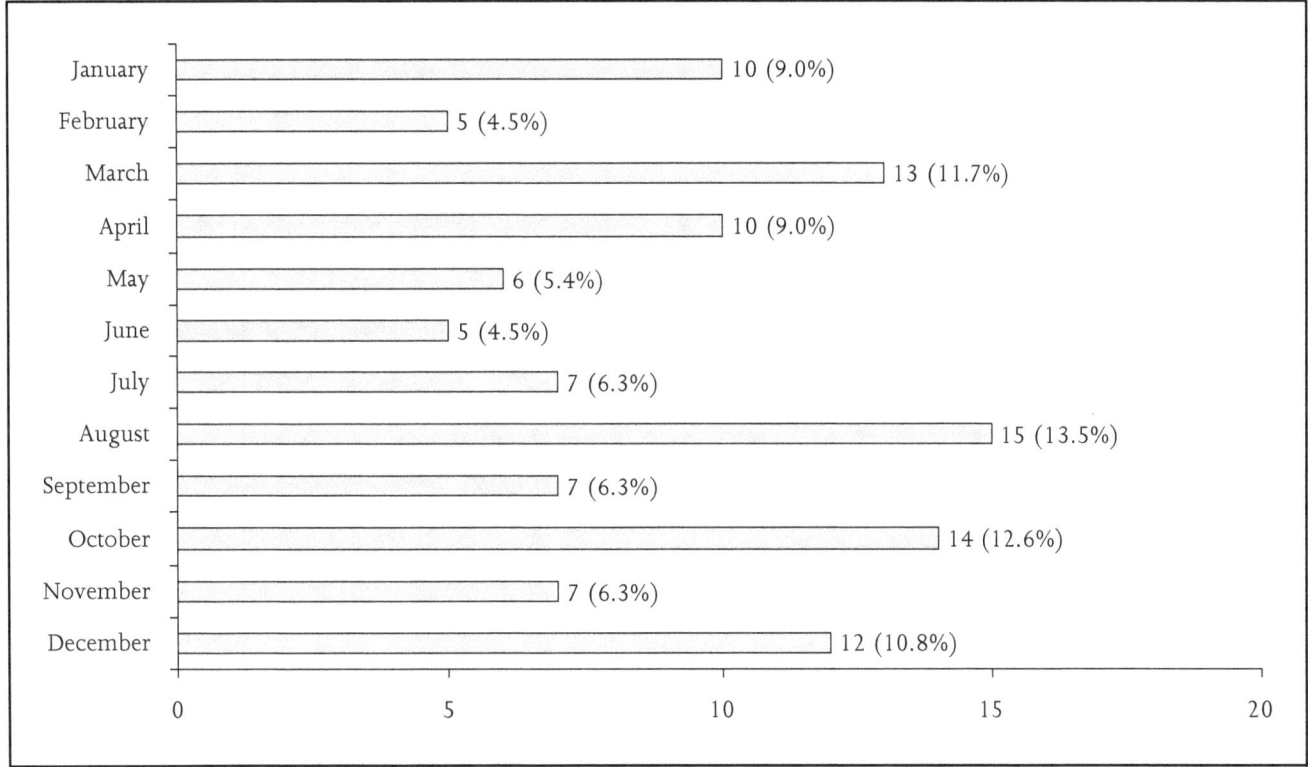

The deadliest month for firefighters was September of 2001 when a total of 355 firefighters died at the World Trade Center and across the county. The next most costly month in the last decade was July of 1994; 21 firefighters died that month including 14 on Storm King Mountain in Colorado.

STATE AND REGION

The distribution of firefighter deaths by State is shown in Table 19*. Firefighters based in 36 states and Guam died in 2003.

The highest number of firefighter deaths based on the location of the fire service organization in 2003 occurred in Oregon with 13 deaths. Eight firefighters died in a van crash, two firefighters died in a helicopter crash, one Oregon-based firefighter died in a helicopter crash in Washington State, one firefighter died in an apparatus crash during training, and one firefighter died of a heart attack while on duty in the fire station.

Table 19. Firefighter Fatalities by State by Location of Fire Service

1	Alabama	0.9%
1	Arkansas	0.9%
3	Arizona	2.7%
3	California	2.7%
3	Connecticut	2.7%
1	Florida	0.9%
3	Georgia	2.7%
1	Guam	0.9%
2	Iowa	1.8%
5	Idaho	4.5%
4	Illinois	3.6%
3	Indiana	2.7%
1	Kansas	0.9%
1	Kentucky	0.9%
5	Louisiana	4.5%
2	Massachusetts	1.8%
2	Maryland	1.8%
1	Maine	0.9%
1	Minnesota	0.9%
2	Missouri	1.8%
1	Mississippi	0.9%
5	North Carolina	4.5%
1	Nebraska	0.9%
4	New Jersey	3.6%
1	New Mexico	0.9%
2	Nevada	1.8%
5	New York	4.5%
4	Ohio	3.6%
13	Oregon	11.7%
10	Pennsylvania	9.0%
4	Tennessee	3.6%
8	Texas	7.2%
1	Vermont	0.9%
1	Washington	0.9%
4	Wisconsin	3.6%
1	West Virginia	0.9%
1	Wyoming	0.9%

Figure 16. Firefighter Fatalities by Region (2003)*

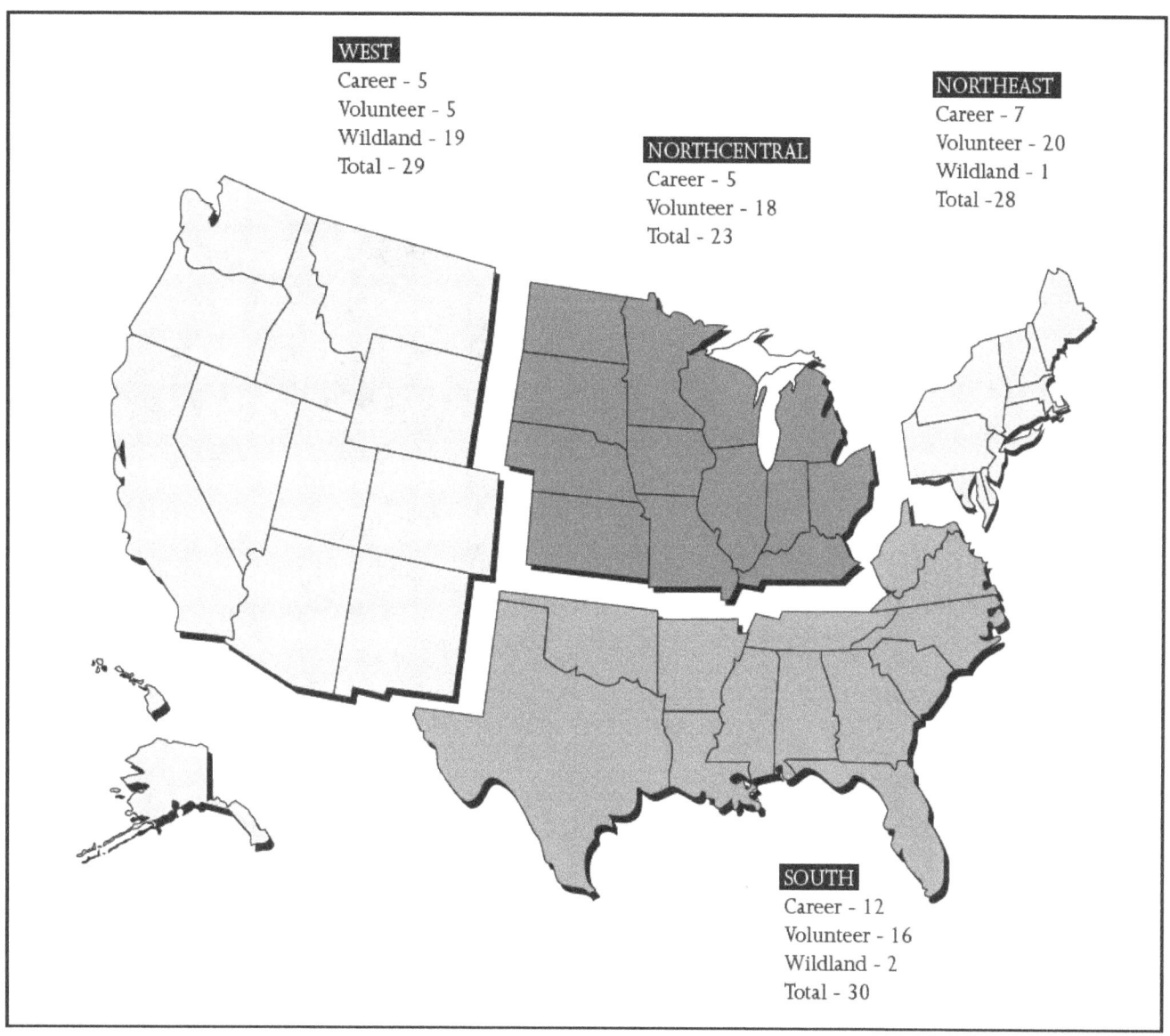

WEST
Career - 5
Volunteer - 5
Wildland - 19
Total - 29

NORTHCENTRAL
Career - 5
Volunteer - 18
Total - 23

NORTHEAST
Career - 7
Volunteer - 20
Wildland - 1
Total - 28

SOUTH
Career - 12
Volunteer - 16
Wildland - 2
Total - 30

* One firefighter fatality occurred in Guam.

45

Figure 17. Firefighter Fatalities By Incident Location

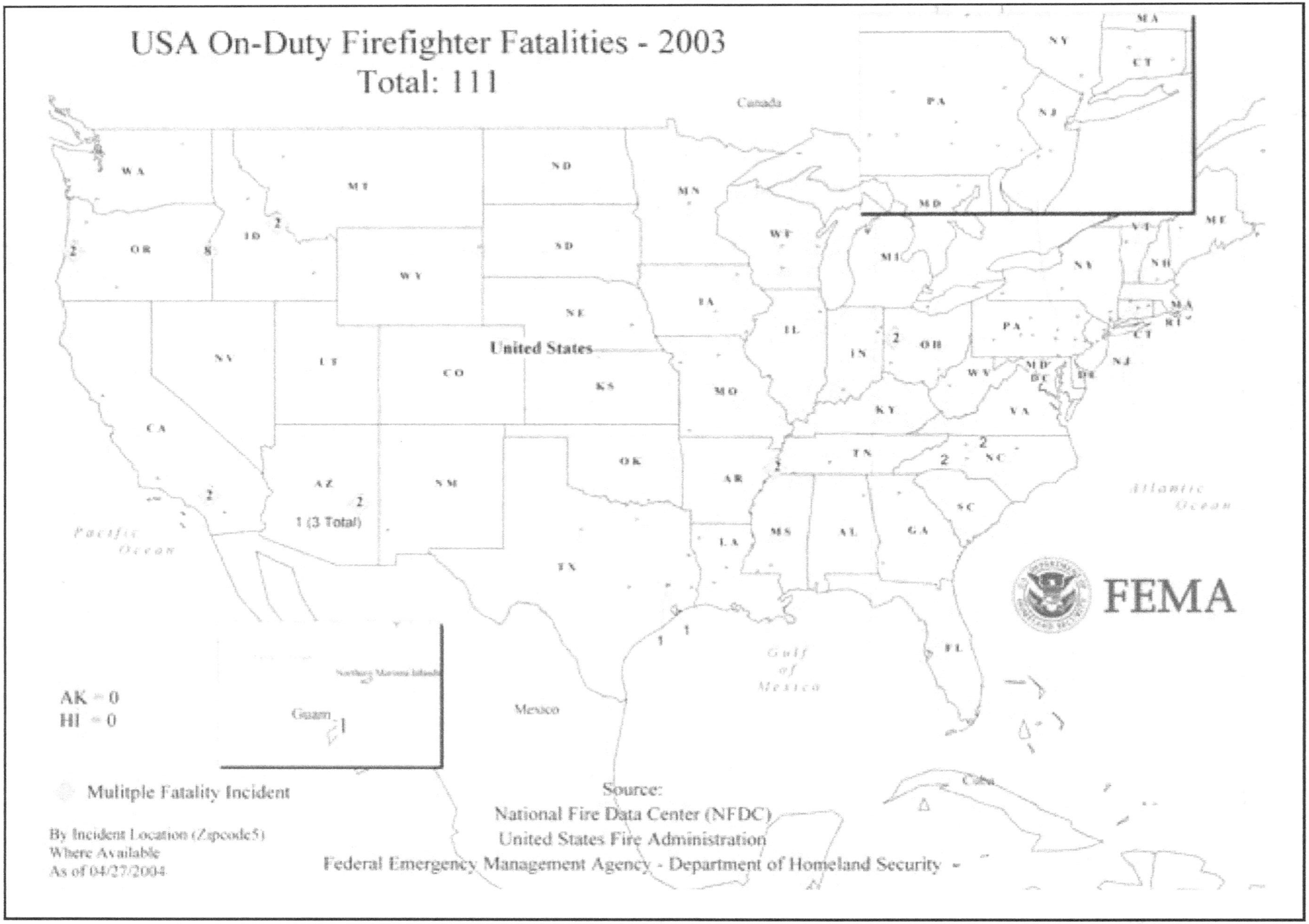
USA On-Duty Firefighter Fatalities - 2003
Total: 111
Canada
United States
Mexico
Pacific Ocean
Atlantic Ocean
Gulf of Mexico
FEMA
1 (3 Total)
Guam
AK = 0
HI = 0
Mulitple Fatality Incident
By Incident Location (Zipcode5)
Where Available
As of 04/27/2004
Source:
National Fire Data Center (NFDC)
United States Fire Administration
Federal Emergency Management Agency - Department of Homeland Security

Figure 18. Firefighter Fatalities By Home Department Location

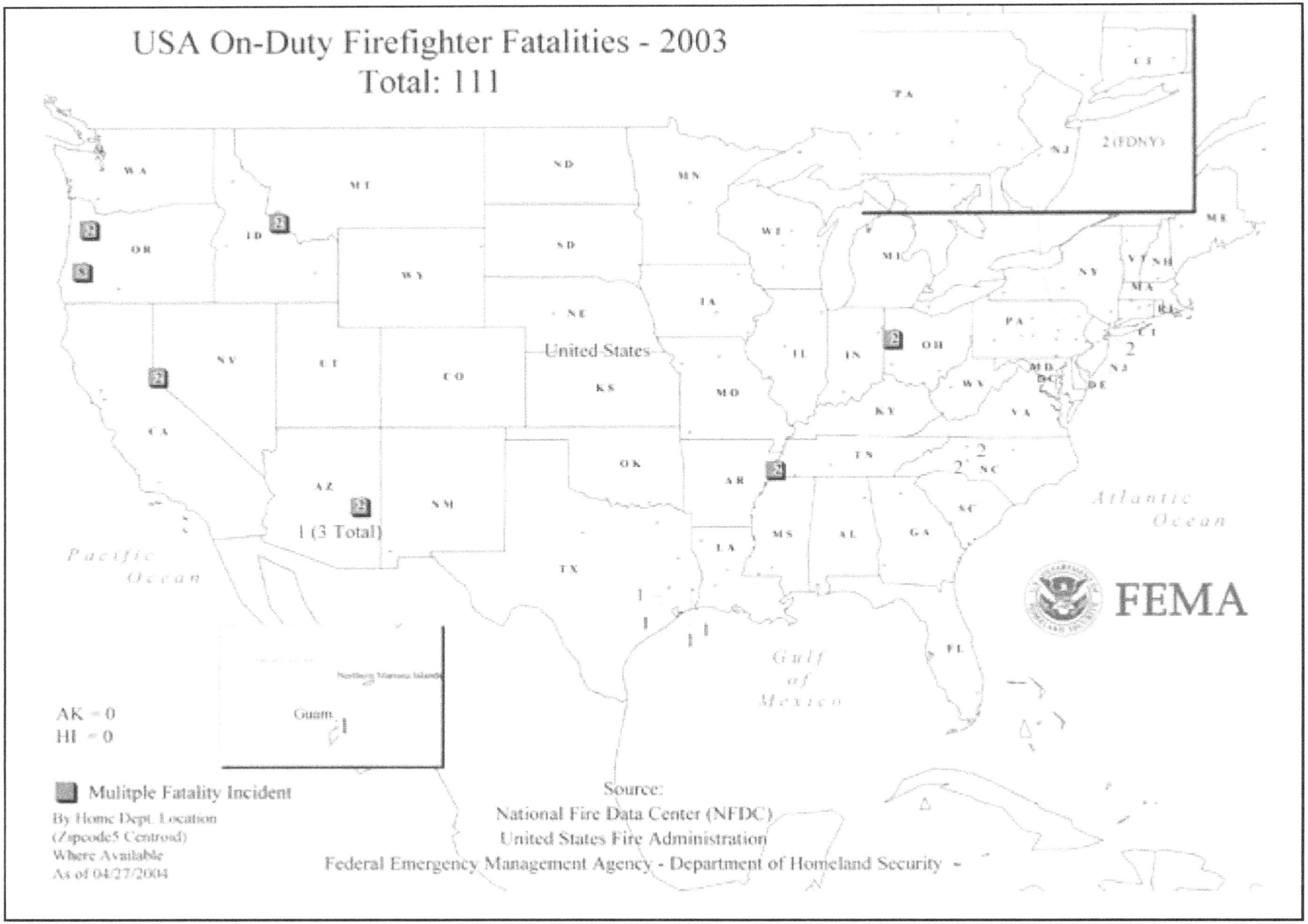

USA On-Duty Firefighter Fatalities - 2003
Total: 111

ANALYSIS OF URBAN/RURAL/ SUBURBAN PATTERNS IN FIREFIGHTER FATALITIES

The United States Bureau of the Census defines *urban* as a place having a population of at least 2,500 or lying within a designated urban area. *Rural* is defined as any community that is not urban. *Suburban* is not a census term but may be taken to refer to any place, urban or rural, that lies within a metropolitan area defined by the Census Bureau, but not within one of the central cities of that metropolitan area.

Fire department areas of responsibility do not always conform to the boundaries used for the census. For example, fire departments organized by counties or special fire protection districts may have both urban and rural coverage areas. In such cases, it may not be possible to characterize the entire coverage area of the fire department as rural or urban, and firefighter deaths were listed as urban or rural based on the particular community or location in which the fatality occurred.

The following patterns were found for 2003 firefighter fatalities. These statistics are based on answers from the fire departments and, when no data from the departments were available, the data are based upon population and area served as reported by the fire departments.

Table 20. Firefighter Fatalities by Coverage Area Type

	Urban/Suburban	Rural	Federal or State Parks/Wildland	Total
Firefighter Fatalities	49	41	21	111

CONCLUSION

The death of any firefighter while on duty is a failure of the firefighter safety system. The failure may be in training, incident command, wellness management, or any of a thousand contributing factors. Each death is a tragedy in the local community and for the entire fire service.

2003 was another year of unacceptable loss for the fire service and the Nation. A trend in the 1990's that revealed a gradual reduction in the number of firefighter fatalities has been reversed with the number of deaths in the past few years at the levels that might have been normal in the 1980's. Indeed, 2003 was a year of unfortunate records:

- The number of wildland firefighter deaths was the highest since 14 firefighters died in Colorado in 1994.

- The number of firefighter fatalities in aircraft-related incidents was the highest since 1994.

- The number of firefighters killed while responding to or returning from incidents was the highest in a decade.

- The number of firefighters killed in vehicle crashes excluding aircraft incidents was the highest since at least 1990.

- The number of deaths due to physical trauma was higher than any year since at least 1990, with the exception of the high toll of September 11, 2001.

This report contains two special topics. The first addresses the issue of alcohol abuse in the general population and the fire service. Methods for identifying those in need of help and getting help for those in need are identified. A firefighter death in 2003 in Wyoming brought this issue to the fore. A policy statement by the International Association of Fire Chiefs (IAFC) provides direction to all firefighters and fire officers.

The second special topic is risk management. Many firefighter fatalities result from poor risk management by firefighters and fire officers. The consistent, everyday use of basic risk management techniques has the potential for saving the lives of firefighters and those we serve.

SPECIAL TOPICS

Alcohol Use in the Fire Service

According to the United States Department of Health and Human Services Alcohol and Drug Information Clearinghouse, approximately 14 million Americans - 7.4 percent of the population - meet the diagnostic criteria for alcohol abuse or alcoholism.

The fire service is not immune from these facts. Alcoholics live in every walk of life in our society, including the fire service. Through their jobs, firefighters see first-hand the effects of alcohol abuse. Firefighters often respond to alcohol-related vehicle crashes, domestic violence assaults, and other situations that are caused or made worse by alcohol.

We have no difficulty seeing the negative impact of alcohol abuse in these situations.

These problems almost always tend to have a direct impact on someone who is "outside" of the fire service. The people who are helped by firefighters when they are involved in an alcohol-related situation are not often part of the fire service family. The facts, however, are at odds with that opinion.

From 1990 through 2003, there were 17 firefighter fatalities where alcohol or drugs were a direct factor in the death of a firefighter – the firefighter who died was drunk or high, or another firefighter involved in the death was drunk or high. Until 2003, alcohol problems in the fire service, when they were talked about at all, always seemed to exist in large urban fire departments.

In 2003, an incident occurred that brought the issue of fire service alcohol abuse to the surface. The incident occurred on May 22nd:

Excerpted from WFS Quarterly Spring/Summer, 2003 Author Unknown

Wyoming student firefighter Anndee Huber, a 16-year-old who was president of her 10th grade class at Newcastle High School, was killed on May 22, 2003, when the fire department tanker she was riding in crashed and rolled. Huber was part of an Explorer program with the Newcastle, Wyoming, Volunteer Fire Department that allowed qualified students to respond to fire calls and help with exterior tasks. She and the driver of the tanker were enroute to a grass fire when the crash occurred.

Anndee Huber was the youngest of four children, and entered the Explorer program as soon as she turned 16. She was also a 4.0 student, and active in swimming and cross-country running. Her brother, Kevin Huber, told reporters, "She was the greatest. How many 16-year-olds do you know who want to be firefighters? Rather than doing the things teenagers do, she was down at the fire hall getting her training." Her oldest brother, Hayden, who also had been a firefighter, learned of his sister's death as he arrived back in the U.S. after 6 months in the Navy in Iraq.

The driver of the tanker involved in the crash was arrested and charged with driving under the influence of alcohol, and with aggravated homicide by vehicle. He was reported to have a history of alcohol problems, including a jail term in 2002 for drunken driving. A police investigation indicated that the driver had been in a local bar up until 15 minutes before the call came in, and had consumed 5 20-ounce beers in approximately 1-1/2 hours.

All firefighters deaths are tragedies. It is often easy to dismiss the details of a firefighter fatality and convince oneself that the situation could never happen in your local fire department. The death of Anndee Huber was an exception to that rule. The death of this young firefighter drew immediate attention to the issue of alcohol abuse in the fire service.

This incident did not involve drinking by career firefighters in a fire station far away in a big city. This incident had as its consequence the death of a bright and dedicated young woman. This tragedy was widely reported in the fire service media.

This incident is a call to action for the fire service to recognize those within the fire service with alcohol abuse problems. Provide them with the help that they need, and insulate the emergency services part of the fire department from the social side of the fire department.

Alcohol and the Fire Service

Many fire departments, primarily volunteer and combination departments, serve as the social center of their communities. The fire station often encompasses a social area that may include facilities for the consumption of alcohol. In addition to the social opportunities that these facilities provide, the fire departments often derive funding from the social operations to provide emergency services. Many fire departments count on the revenue generated from fund drives and social events to provide funding for the fire department.

The IAFC called for its members to adopt the zero-tolerance alcohol policy. The IAFC recognized that many fire departments raise funds through social operations and did not call for that practice to end. The IAFC called for separation of those facilities to help ensure the application of the zero-tolerance standard. The impact of the IAFC policy is not limited to volunteer fire departments or those with social facilities. The 8-hour time between alcohol consumption and work affects all firefighters, including career firefighters.

The challenge is to find a balance between the benefits of the social operations of the fire department and the emergency services provided by the department. The International Association of Fire Chiefs (IAFC) has provided some guidance on this issue.

Zero Tolerance for Alcohol and Drinking in the Fire and Emergency Service

This policy statement is most easily described as a "zero-tolerance" standard about the use of alcohol by members of any fire or emergency services agency/organization at any time when they may be called upon to act or respond as a member of those departments.

Basically, if someone has consumed alcohol within the previous eight (8) hours, or is still noticeably impaired by alcohol consumed previous to the eight (8) hours, they must voluntarily remove themselves from the activities and functions of the fire or emergency services agency/organization, including all emergency operations and training.

No member of a fire & emergency services agency/organization shall participate in any aspect of the organization and operation of the fire or emergency agency/organization under the influence of alcohol, including but not limited to any fire and emergency operations, fire-police, training, etc.

No alcohol shall be on the premises of any operational portion of the fire department, including but not limited to the apparatus, the apparatus floor, the station living areas, etc.

Fire & emergency services agencies/organizations which raise funds by operating and/or renting social halls must provide a clear and distinct separation of facilities to help insure the zero-tolerance standard of alcohol consumption by their members who may be called upon to perform official duties.

All fire & emergency service agencies/organizations should develop written policies and have procedures in place to support and enforce this policy recommendation. Included in such a policy should be provisions for blood alcohol testing of any individuals involved with any accident that causes measurable damage to apparatus or property or injury/death of agency/organization personnel or civilians.

How to Recognize Alcohol Abuse

The use of alcohol in a social setting is accepted by many Americans. Alcohol abuse can be difficult to discern from the controlled use of alcohol.

Answering the following four questions can help you find out if you or someone you know has a drinking problem:

- Have you ever felt you should cut down on your drinking?

- Have people annoyed you by criticizing your drinking?

- Have you ever felt bad or guilty about your drinking?

- Have you ever had a drink first thing in the morning to steady your nerves or to get rid of a hangover?

One "yes" answer suggests a possible alcohol problem. More than one "yes" answer means it is highly likely that a problem exists.

Most alcohol abusers will need help to recover from alcoholism. It is very difficult for alcoholics to eliminate their dependence on alcohol by themselves. There are a number of programs that are available at the local level to help.

How to Help

Firefighters and fire officers can help eliminate the presence of alcohol in the fire service in many ways. The principal means of fighting this problem are

- Maintain an awareness of the signs of alcoholism.

- Adopt a written policy for each fire department that deals with the use of alcohol and the eligibility of firefighters to respond to emergencies and participate in fire department activities. Most importantly, it is everyone's duty to support and enforce the policy when adopted. The words on paper will not mean anything unless fire officers and individual firefighters make the policy work in the real world.

- Help those with alcohol abuse problems seek help.

Alcoholics Anonymous was founded in 1935. Alcoholics Anonymous (AA) is an international fellowship of men and women who have had a drinking problem. It is nonprofessional, self-supporting, multiracial, apolitical, and available almost everywhere. There are no age or education requirements. Membership is open to anyone who wants to do something about his or her drinking problem. AA can be contacted at www.alcoholics-anonymous.org. The site contains a link for finding a local meeting.

There is not sufficient information in this venue to provide comprehensive guidance for those that seek to help someone with an alcohol abuse problem. There are many resources available at the local level to help. The key is recognizing the problem and then making treatment resources available.

Risk Management

This section presents information on risk management. The classic risk management method is presented and then followed by application of risk management techniques to nonemergency, pre-emergency, and emergency risk management examples.

NFPA 1500, *Standard on Fire Department Occupational Safety and Health Program*, requires the development and adoption of an organizational risk management plan. NFPA 1500 provides the following definitions:

Risk – A measure of the probability and severity of adverse effects that result from an exposure to a hazard.

Risk Management – The process of planning, organizing, directing, and controlling the resources and activities of an organization in order to minimize detrimental effects on that organization.

NFPA 1500 requires the use of four risk management components. These components make up the classic risk management model.

The following section, adapted from the National Fire Academy *Incident Safety Officer* curriculum, illustrates the application of the classic risk management method.

The classic risk management model presents a systematic approach for identifying and planning for the control of risks. This methodical process for making decisions can be used not only for the nonemergency risks that all organizations must address,

> The United States Fire Administration publication "Risk Management Practices in the Fire Service" is available free of charge from the USFA Publications Center. The publication, FA-166, is available at www.usfa.fema.gov under the Publications section.

but also for the risks associated with the response to and mitigation of an emergency incident. The factors at each incident will always vary, but the classic risk management model applies equally to all situations.

The model used by NFPA 1500 has four primary components, or steps, which serve as a foundation for this process. Each one depends upon information generated by the previous step, so it is important to evaluate each one before moving on to the next. These four steps are discussed in detail in the following sections.

- Risk identification

- Risk evaluation

- Risk control techniques

- Monitoring

Risk Identification

What might go wrong? Compile a list of all emergency and nonemergency operations in which the department participates. Ideally, plan for the worst, but hope for the best. There are many sources to assist with this identification process. The first, and possibly the most effective, is the department's loss prevention data. Seek input and ideas from personnel, trade journals, professional associations, and other service providers.

When using ideas from other fire departments or organizations, simply consider local circumstances when formulating the list.

Risk Evaluation

Once the risks are identified, they can be evaluated from both a frequency and a severity standpoint. Frequency addresses the likelihood of occurrence. Typically, if a particular type of incident (e.g., back injuries) has occurred repeatedly, these incidents will continue to occur until effective control measures are implemented.

Severity addresses the degree of seriousness of the incident. This can be measured in a variety of ways, such as time away from work, seriousness of injuries that can result from an activity, cost of damage, cost of (and time for) repair or replacement, disruption of services, and legal costs. Using the information gathered in the identification step, the risks can be classified based on severity.

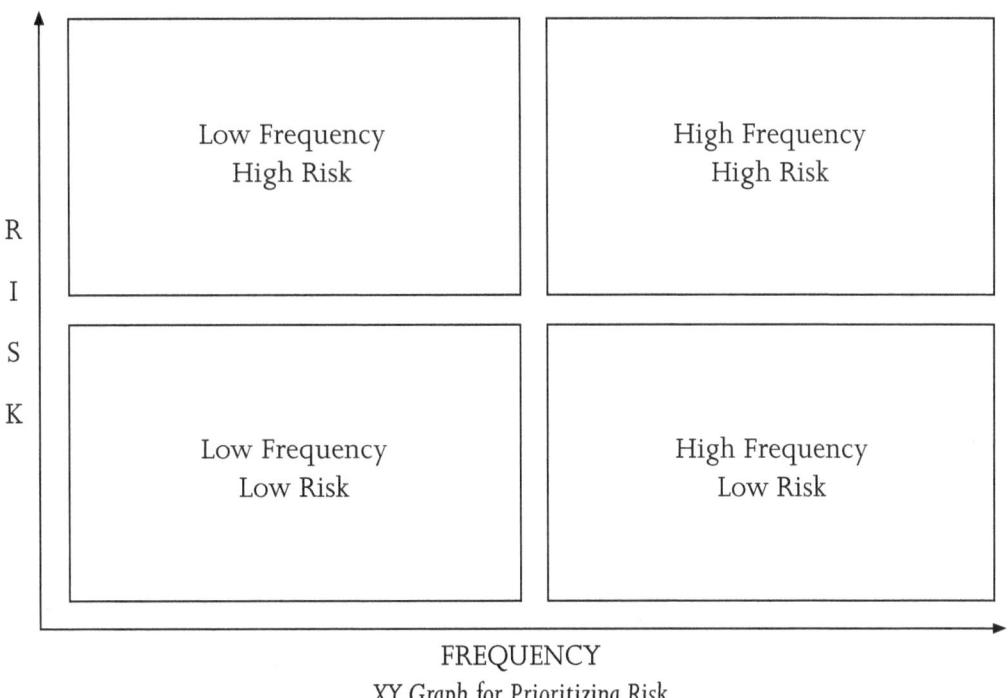

FREQUENCY

XY Graph for Prioritizing Risk

Taken in combination, the results of the frequency and severity determinations will help to establish priorities for determining action. Any risk that has a high probability of occurrence and will have serious consequences (high risk or severity), deserves immediate action, and would be considered a high-priority item. Nonserious incidents with a low likelihood of occurrence are a lower priority and can be placed near the bottom of the "action required" list.

Risk Control Techniques

At this point in the process, risks have been identified and evaluated, so it is time to find solutions. There are several approaches to take, including risk avoidance, implementation of control measures, and risk transfer.

In any situation, the best choice is risk avoidance. Simply put, this means avoid the activity that creates the risk. In an emergency services organization, this frequently is impractical. Lifting a stretcher presents a serious back injury risk, but you cannot avoid this risk and still provide effective service.

An example of where avoidance has been very practical is the widespread use of sharps containers. The risks associated with recapping needles are well documented; therefore, recapping is no longer an accepted practice. This risky behavior can be avoided through the proper use of a sharps container.

The most common method used for the management of risk is the adoption of effective control measures. While control measures will not eliminate the risk, they can reduce the likelihood of occurrence or mitigate the severity. Safety programs, ongoing training and education programs, and well-defined Standard Operating Procedures (SOP's) are all effective control measures.

Some typical control measures instituted to control fireground injuries include accountability, use of full protective clothing, a mandatory respiratory protection program, training and education, and competent SOP's. These control measures coupled together make an effective program that ensures safe fireground operations.

Risk transfer can be accomplished in two primary ways: physically transferring the risk to somebody else or through the purchase of insurance. For a fire or EMS organization, the transfer of risk may be difficult if not impossible. As an example, some emergency service organizations have transferred part of the risk of hazardous materials incident response with the use of private contractors to clean up scenes after the incident is controlled.

The purchase of insurance transfers financial risk only. In addition, it does nothing to affect the likelihood of occurrence. Buying fire insurance on a fire station, while highly recommended to protect the assets of the department, does nothing to prevent the station from burning down. Therefore, insurance is no substitute for effective control measures.

Risk Monitoring

The last step in the process is risk management monitoring. Once control measures have been implemented, they need to be evaluated to measure their effectiveness. Any problems that occur in the process have to be revised or modified. This final step ensures that the system is dynamic and will facilitate periodic reviews of the entire program.

The intent of the risk management plan is to develop a strategy for reducing the inherent risks associated with fire department operations. Regardless of the size or type of fire department, every organization should operate within the parameters of a risk management plan. This is a dynamic and aggressive process that must be monitored and revised annually by the health and safety officer.

Risk Management Model Application

Risk management techniques may be applied to every part of the operation of a fire department to improve the level of safety and occupational health for fire department members. Limiting the damage and cost that result from on the job accidents is an activity that is best and most effectively done before the emergency occurs.

The following sections will discuss the application of the classic risk management model to nonemergency, pre-emergency, and emergency situations. Examples will also be provided.

Nonemergency Risk Management

There are risks to firefighter safety and health that have very little to do with the response to an emergency. Fire department facilities are subject to the same risks as any building so appropriate steps to secure the workplace should be taken. These steps can include the provision of automatic fire sprinklers, smoke detection systems, proper emergency exiting, nonslip flooring, carbon monoxide detectors, and natural gas detectors.

The assurance of safe and effective fire department apparatus and equipment is another area of nonemergency risk management. The provision of vehicles that comply with national standards, as a minimum, will address a number of potential risks. Vehicles should be regularly maintained, have braking systems that are designed for emergency response, and complete required pump and/or aerial ladder service tests.

> ### Diesel Exhaust Exposure
> Exposure to diesel exhaust can cause a number of temporary symptoms including headaches and respiratory problems. NIOSH has determined that diesel exhaust is a potential carcinogen for humans and long-term effects are possible.
>
> Nonemergency risk management measures to avoid exposure to diesel exhaust can include exhaust capture systems, a reduction in vehicle indoor running time, duct cleaning, maintenance of a positive pressure in fire station living areas, and delays in door closure after apparatus returns to the station.

Risk management techniques also should be applied to training. The assignment of a safety officer to burn drills, the use of an appropriate instructor-to-student ratio, and the use of NFPA 1403 for fires in acquired structures will improve the level of safety for firefighters during training.

Pre-Emergency Risk Management

Pre-emergency risk management uses the classic risk management model approach. Risks that can be managed in advance are identified. This is the process that occurs prior to the response to emergencies, but which will make emergency scene risk management easier to perform. Pre-emergency risks are those that fit between the nonemergency risks and the risks presented at an emergency incident.

The pre-emergency risk management elements must be identified and managed in order for an organization to conduct emergency operations safely and effectively. Therefore, pre-emergency risk management can be defined as "a process that utilizes key safety and health elements, prior to response, that will reduce risks involved during emergency operations and enhance customer service."

> ### United States Firefighter Disorientation Study
>
> A report prepared by Captain William Mora of the San Antonio Fire Department analyzed 17 incidents where firefighters became disoriented in structures that resulted in a total of 23 firefighter fatalities.
>
> The study reveled a disorientation sequence common to all of these incidents that included light smoke showing upon arrival, an aggressive interior attack, deteriorating interior conditions, and firefighters becoming separated from handlines.
>
> The study recommends SOP's and training that includes a cautious initial assessment and a managed initial attack, if warranted.
>
> The full study is available at:
> www.sanantonio.gov/safd/pdf/FirefighterDisorientationStudy.pdf

Pre-emergency risk management can include programs as diverse as Incident Command System (ICS) training, the purchase and provision of appropriate personal protective equipment (PPE) and clothing such as SCBA's and helmets, and the study of past incidents to learn lessons from others.

Emergency Risk Management

In contrast to the studied approach of nonemergency and pre-emergency risk management, risk management during emergencies is a constantly changing, fast-paced activity. Nonemergency and pre-emergency risk management activities have been completed and will assist with the safety of firefighters on the emergency scene.

Everyone on the emergency scene has a safety and risk management responsibility. Standards require the presence of safety officers at many incident scenes; however, safety officers cannot be everywhere and cannot be on every incident scene.

Risk versus Benefit

Every firefighter and fire officer should view every activity related to an emergency through a risk/benefit "filter." This risk/benefit filtering begins when the alarm sounds and continues until all firefighters have returned safely from the emergency. The filter is really a shortened version of the department's risk management program and appears in NFPA 1500:

1. Activities that present a significant risk to the safety of members shall be limited to situations where there is a potential to save endangered lives.

2. Activities that are routinely employed to protect property shall be recognized as inherent risks to the safety of members, and actions shall be taken to reduce or avoid these risks.

3. No risk to the safety of members shall be acceptable when there is no possibility to save lives or property.

This section of NFPA 1500 can be summarized as "take calculated risks to save a life, take small risks to save property, risk nothing to save what is already lost."

The risk/benefit filter can be used constantly by every firefighter and fire officer involved in the emergency. The filtering process takes place in less than a second, but its product will help to make sure that firefighters and the customers that receive our service are safe.

Some examples:

- My department has been dispatched to a dumpster fire in an apartment complex. I know from personal observation that the dumpster has no exposures. I choose to respond to my fire station at moderate speed, taking no extra risks, since a faster response time to the dumpster will have no benefit to the outcome of the incident.

- Just before sunset, my department is dispatched to a water rescue in a local stream. A boater is missing and bystanders suspect the boater has been trapped under a tree in the middle of the fast-moving stream for 45 minutes. As the incident commander of the scene, I do not permit firefighters to enter the water. In my judgment, the boater is certainly dead and no benefit will come from the risk of placing firefighters in the water at this time.

- My engine company arrives on the scene of a house fire in the middle of the night. The driveway is filled with cars and neighbors tell us that a family is still inside. Smoke is showing from the windows and a glow can be seen in windows at the rear of the structure. As the company officer, I order an attack line to the front door and join my crew for a search of the interior. In my judgment, the deployment of my crew to the interior was worth the risk based on the reports of trapped occupants.

- My engine company is dispatched to an emergency medical incident in a private home. Before I begin to treat the patient, I put on examination gloves. In my opinion, the benefit of avoiding disease is worth the risk of offending the patient.

- My fire department is dispatched to a fire in a large residence that is far from any water supply. Upon our arrival on the scene, we find a well-involved structure fire and learn from the residents that everyone is out of the house and accounted for. As the incident commander, I order an exterior attack on the fire until a reliable water supply can be established. In my judgment, the risk of entering the structure with limited water was not worth the benefit of what we might be able to save.

- My department and a number of mutual-aid fire departments have been fighting a fire in a single-story warehouse for hours. Ladder pipes and deck guns have been used to control the main body of the fire, yet smoldering piles of stock remain. As the incident commander, I choose not to deploy firefighters to the interior for overhaul. The risk of collapse is not worth the benefit of complete fire extinguishment.

If the risk/benefit filter was used by every firefighter and every fire officer on each emergency, the result would likely be an immediate reduction in the tragic loss of life that this report details. The people that are served by the fire service expect that firefighters will take risks that are unacceptable to the general public; training and equipment allow firefighters to accept some risks. Firefighters are not expected to give their lives in vain to save what is already lost.

Cincinnati Fire Chief Robert Wright carries the flag that covered Firefighter Oscar Armstrong's casket as he prepares to present it to Armstrong's mother.

APPENDIX
SUMMARY OF 2003 INCIDENTS

January 8, 2003 – 1050hrs
Lattie Floyd Collins, III, Firefighter
Age 36, Volunteer
Donalsonville Fire Department, Georgia

Firefighter Collins was responding in his personal vehicle to a report of a motor vehicle crash. As he neared the fire station, he came to an intersection and met a responding engine approaching from the opposite direction. Firefighter Collins' vehicle and the engine stopped at their respective stops signs. Firefighter Collins was preparing to proceed across the intersection en route to the fire station and the engine was preparing to make a turn.

As Firefighter Collins proceeded through the intersection, his vehicle was struck on the right side by a vehicle traveling on the intersecting highway. The force of the collision drove Firefighter Collins' vehicle through the intersection, almost striking the responding engine. Firefighter Collins had apparently not seen the vehicle that struck him due to a blind spot or as a result of being focused on the presence of the engine.

Firefighters immediately went to the aid of Firefighter Collins and found him lying across the front-seat of his car with his head against the passenger's door. Firefighters could not find any vital signs but attempted to maintain an airway as extrication was accomplished.

Firefighter Collins was transported by ambulance to the hospital where he was pronounced dead upon arrival. The cause of death was listed as closed head trauma complicated by positional asphyxia. Firefighter Collins was not wearing a seatbelt at the time of the crash. Firefighter Collins had been a member of the Donalsonville Fire Department for 3 months.

January 8, 2003 – 1719hrs
Melinda Jean "Mindy" Ohler, Firefighter
Age 46, Career
San Francisco Fire Department, California

Firefighter Ohler was a passenger in an engine company responding to a fire alarm. Firefighter Ohler was riding in the open jump seat of the reserve apparatus behind the driver. Prior to moving the apparatus, the officer noted that both firefighters were seated and received a hand signal and a verbal indication from Firefighter Ohler that he could begin the response.

As the apparatus traveled through a right-hand curve on a freeway on-ramp, Firefighter Ohler apparently unbuckled her seatbelt to reach for her hearing protection. As the apparatus turned, she lost her balance and fell from the apparatus. She struck her head on the pavement and sustained a serious injury.

Firefighter Ohler was transported by fire department ambulance to the hospital for treatment. She remained in a coma until her death on January 13, 2003.

The San Francisco Fire Department was fined by CAL-OSHA for failure to have a safety rail installed on the apparatus involved in Firefighter Ohler's death, as well as other apparatus of the same vintage. The fire department also was directed to enforce the use of seatbelts.

January 9, 2003 – 1932hrs
James A. "Matt" Pereira, Firefighter I
Age 40, Career
Guam Fire Department, Agat Fire Station, Guam

Firefighter Pereira had returned to quarters with the members of his fire company after responding to an EMS incident and an unauthorized burning incident. Firefighter Pereira was resting. Other firefighters heard him emit a snoring sound and attempted to wake him. He was unconscious, unresponsive, and experiencing convulsions.

CPR was initiated and an AED was applied. ALS and an ambulance were called and Firefighter Pereira was transported to the hospital. He was pronounced dead upon arrival.

The cause of death was listed as pulmonary edema following smoke exposure.

January 14, 2003 – 0129hrs
Joseph Michael Rotherham, Captain
Age 46, Career
Springfield Fire/Rescue Department, Illinois

Captain Rotherham and the members of his engine company responded with other companies to the report of a structure fire in a 1-1/2 story duplex. The first fire department unit on the scene, a battalion chief, reported heavy smoke.

Upon their arrival, Captain Rotherham and his crew were assigned to search the building and attempt to locate the fire. Captain Rotherham was wearing full structural firefighting protective clothing and equipment, including a SCBA. The fire was located in the attic and Captain Rotherham and his crew stretched an attack line into the building.

After ordering the line to be charged, Captain Rotherham reentered the structure and began to advance the line toward the fire. He became disoriented and handed the nozzle to his firefighter. After reaching the second floor and applying water to the fire, the Captain ordered his firefighter to abandon the line and leave the structure. Captain Rotherham stated that he was running out of air.

While sitting on his apparatus having his SCBA cylinder changed, Captain Rotherham suddenly collapsed. Firefighters began CPR immediately and Captain Rotherham was loaded into an on-scene ambulance. Advanced Life Support (ALS) procedures were provided in the ambulance and continued in a hospital emergency room. Despite these efforts, Captain Rotherham was pronounced dead at the hospital.

The cause of death was listed as a cardiac arrhythmia due to heart disease.

January 17, 2003 – 2256hrs
James W. McAuley, Fire Police Officer
Age 78, Volunteer
Polk Township Volunteer Fire Department, Pennsylvania

Fire Police Officer McAuley and the members of his department were dispatched to respond to a car fire on a local highway. Fire Police Officer McAuley assisted with traffic control and then waited in his vehicle for further orders.

After the fire was extinguished, a fire truck left the scene to refill its water tank and passed the spot where Fire Police Officer McAuley's vehicle was staged. As the apparatus passed, Fire Police Officer McAuley moved his vehicle to the side of the road.

Approximately 5 minutes later, another fire truck left the scene and found Fire Police Officer McAuley slumped over the steering wheel of his vehicle. A firefighter removed Fire Police Officer McAuley from his vehicle and started CPR. An AED was used and administered two shocks. An ambulance arrived and transported Fire Police Officer McAuley to the hospital where he was pronounced dead.

The cause of death was listed as a heart attack.

January 19, 2003 – 0645hrs
James Edward Taylor, Firefighter/EMT-I
Age 28, Career
Bonham Fire/Rescue, Texas

Firefighter Taylor was the front-seat passenger in a 1996 Ford F350 Type III ambulance that was responding to a motor vehicle crash. Firefighter Taylor was wearing a seatbelt. The ambulance was traveling at a speed of approximately 75 miles per hour. As the ambulance crested a hill on a two-lane road, it was met head-on by an oncoming passenger vehicle. After the crash, both vehicles caught fire.

Firefighter Taylor was trapped in the vehicle and the driver of the ambulance was injured. The ambulance driver was able to get out of the vehicle and call for assistance. The driver attempted to control the fire with an extinguisher but was unsuccessful due to the intensity of the fire. Arriving firefighters controlled the fire; Firefighter Taylor was pronounced dead at the scene.

The driver of the passenger car also was killed. The cause of the crash was determined to be driver inattention on the part of the driver of the passenger car. The passenger car crossed the center line and struck the responding ambulance.

The cause of death for Firefighter Taylor was listed as blunt force injuries.

The Texas State Fire Marshal has prepared a thorough report on this incident. The report is available at www.tdi.state.tx.us/fire/fmloddinvesti.html

For additional information regarding this incident, please refer to NIOSH Fire Fighter Fatality Investigation and Prevention Program report F2003-05 (www.cdc.gov/niosh/face200305.html).

January 19, 2003 – 0945hrs
Gary L. Staley, Firefighter
Age 33, Volunteer
Porter Volunteer Fire Department, Texas

Firefighter Staley and members of his department and mutual-aid departments were dispatched to a fire in a commercial occupancy. The fire involved a specialized vehicle restoration shop.

The first-arriving command officer reported a working fire and requested additional resources. Firefighter Staley arrived on the first engine company. Firefighter Staley and other firefighters stretched a 1-3/4-inch handline with Compressed Air Foam (CAF). Upon entry to the shop, they found light smoke from the ceiling to the floor. A portable positive pressure fan was placed in a doorway.

The fire began to intensify and began to roll over the top of the attack crew. One attack crewmember was forced to leave because his hands were burning. As this firefighter reached the exterior, air horns began to sound indicating that firefighters should evacuate the building. Two of the attack line crewmembers exited the building – both suffered burns. Firefighter Staley did not exit.

Approximately 30 seconds after the third firefighter left the building, an explosion occurred inside the building.

Firefighters quickly determined that Firefighter Staley was missing. A Rapid Intervention Crew (RIC) was deployed to the interior but they were forced to exit the building when evacuation horns were sounded again. Ladder pipes were used to knock down the fire and the RIC was allowed to reenter the structure. The RIC was forced to leave again due to fire progress and an unreliable water supply. After master streams were used to knock down the fire again, firefighters located Firefighter Staley inside the structure.

Firefighter Staley likely became disoriented inside the structure and ran out of air. He received injuries from the explosion, both of his eardrums were ruptured and there was damage to his lungs consistent with the explosion. He was found in a prone position with the regulator removed from his SCBA facepiece. His PASS device was sounding but was muffled by the position of his body.

The cause of death was listed as thermal injuries with smoke inhalation and blast injuries.

The fire began when a flooring contractor used a flammable liquid to prepare a floor. Vapors were ignited when a portable heater was used to speed the drying process.

For additional information regarding this incident, please refer to NIOSH Fire Fighter Fatality Investigation and Prevention Program report F2003-03 (www.cdc.gov/niosh/face200303.html).

January 20, 2003 – 1700hrs
Keith Robert Hess, Firefighter/EMT
Age 22, Career
Fannett-Metal Fire & Ambulance Company, Pennsylvania

Firefighter Hess responded with the members of his department to a fire in a large residence. After the main body of the fire was controlled, a process that took about 2 hours, firefighters moved into the structure to locate and extinguish hot spots.

A number of firefighters were in the structure, including Firefighter Hess. A centrally located chimney collapsed and brought down the second floor of the structure onto firefighters. Three firefighters were trapped by the collapse; one was able to extricate himself and get out of the structure. Firefighters entered the building and located the two trapped firefighters. One firefighter was removed and suffered minor injuries. Firefighter Hess was severely injured and he was found to be without a pulse and not breathing.

Firefighter Hess was transported by helicopter to a local hospital where he was pronounced dead. The cause of death was listed as compressional asphyxia.

Firefighter Hess was also a Lieutenant with the West End Fire and Rescue Company.

January 21, 2003 – 2030hrs
Dennis G. Mignerone, Captain
Age 50, Career
Webster Groves Fire Department, Missouri

Captain Mignerone was participating in on-duty physical fitness exercises, including weightlifting and walking on a treadmill. He began to experience chest pains and alerted other crew members.

The crew began treatment for angina and transferred Captain Mignerone to a fire department ambulance when his condition did not improve. While in the ambulance, he went into cardiac arrest. CPR and ALS procedures were initiated in the ambulance and continued upon arrival at the emergency room.

Despite efforts to save him, Captain Mignerone was pronounced dead in the hospital after resuscitation efforts were discontinued.

The death certificate listed the cause of death as acute myocardial infarction due to atherosclerotic coronary artery disease.

For additional information regarding this incident, please refer to NIOSH Fire Fighter Fatality Investigation and Prevention Program report F2003-24 (www.cdc.gov/niosh/face200324.html).

January 25, 2003 – 1330hrs
Michael Wayne Copeland, Captain
Age 50, Career
Charlotte Fire Department, North Carolina

Captain Copeland and his engine company were participating in a multicompany live structural fire training exercise. Captain Copeland's engine was to act as the first unit on the scene of the exercise.

At the start of the training activity, Captain Copeland's engine laid a supply line into the scene. Captain Copeland took command of the incident from his position in the cab of the apparatus, gave an arrival report, and assigned tasks to other companies. Captain Copeland dismounted the engine and donned his SCBA. He proceeded to the rear of the truck. He was observed by other firefighters bending over with his hands on his knees. He then sat down on the back step of the engine and began to remove his facepiece. He told a battalion chief that he could not catch his breath. As firefighters removed his protective clothing, Captain Copeland slumped into their arms.

An ambulance was called and firefighters began to treat Captain Copeland. CPR was begun and Captain Copeland was lifted into the ambulance upon its arrival. ALS procedures were initiated in the ambulance and continued in the hospital emergency room. Despite all efforts, Captain Copeland was pronounced dead at the hospital.

Captain Copeland had a history of heart problems but was medically certified as eligible for full duty in December of 2002. The cause of death was listed as an acute myocardial infarction due to coronary atherosclerosis disease.

The Charlotte Fire Department suffered a second firefighter fatality in April of 2003.

February 1, 2003 – 0901hrs
Robert John Moseley, Engineer
Age 52, Career
Santa Barbara County Fire Department, California

Engineer Moseley and the members of his crew were working to clear a trail on a steep hill behind the fire station. The firefighters were using a lawn mower to accomplish the task; the trail is about 700 feet in length at a 20 percent grade.

Engineer Moseley was wearing web gear with a rope attached from his back to the lawn mower. While Engineer Moseley pulled, another firefighter pushed the lawn mower. After they reached the top of the hill, Engineer Moseley and the other firefighter stopped for a rest. Everything seemed fine.

The other firefighter began to descend the hill mowing a wider path. Engineer Moseley followed the firefighter down the hill. At some point, the firefighter looked behind him and observed Engineer Moseley lying down.

The firefighter ran to Engineer Moseley's side and found him unresponsive and not breathing. The firefighter called for help on the radio and began CPR. An AED was attached and delivered two shocks. Paramedics arrived and continued care during the trip to the hospital. Engineer Moseley was pronounced dead at the hospital.

The cause of death was listed as atherosclerotic coronary artery disease.

Santa Barbara County Fire Department Web site - www.sbcfire.com

---

February 12, 2003 – 0826hrs
Wayne Kevin Clarke, Firefighter Trainee
Age 46, Career
Dallas Fire-Rescue Department, Texas

Firefighter Trainee Clarke was participating in a highrise drill with other members of his recruit class. The drill involved three trips to the top of a six-story training tower carrying firefighting equipment while wearing full structural protective clothing and an SCBA. On the third and final trip up the stairs, Firefighter Trainee Clarke began to experience trouble at the fourth-floor landing. He completed his trip to the top of the tower; however, he appeared to be extremely fatigued and disoriented.

Firefighter Trainee Clarke began his decent but was unable to continue. Noting his condition, a training officer instructed other recruits to carry Firefighter Trainee Clarke to the exterior. Once outside, his vital signs were taken.

CPR and ventilation were initiated and an ambulance and paramedics were summoned. A cardiac monitor was attached and Firefighter Trainee Clarke was shocked. CPR and ALS protocols were continued in the ambulance and for approximately 20 minutes in the hospital emergency room. All efforts were unsuccessful and Firefighter Trainee Clarke was pronounced dead.

The cause of death was listed as hypertrophy, an enlarged heart.

Dallas Fire-Rescue Web site - www.dallasfirerescue.com

For additional information regarding this incident, please refer to NIOSH Fire Fighter Fatality Investigation and Prevention Program report F2003-21 (www.cdc.gov/niosh/face200321.html).

February 20, 2003 – 0703hrs
Larry Allen Brown, Captain
Age 53, Career
Kingsley Field Fire Department, Oregon

Captain Brown and the members of his crew spent their shift performing normal duties such as fire extinguisher checks and participating in aircraft emergency egress training exercises. Captain Brown was last seen by his crew at approximately 2030hrs as he prepared to go to sleep.

Firefighters found Captain Brown unresponsive laying diagonally across his bed when they went to check on him at approximately 0700hrs the next morning. Paramedics and an ambulance were called and firefighters assessed Captain Brown. Due to the fact that he was obviously deceased, no resuscitative measures were attempted.

The cause of death was listed as arteriosclerotic cardiovascular disease.

For additional information regarding this incident, please refer to NIOSH Fire Fighter Fatality Investigation and Prevention Program report F2003-26 (www.cdc.gov/niosh/face200326.html).

Kingsley Field Fire Department Web site - www.kingsleyfieldfire.com

---

February 23, 2003 – 0911hrs
Woodrow W. "Woody" Pinkerton, Fire Police Captain
Age 63, Volunteer
Medford Division of Fire, New Jersey

Fire Police Captain Pinkerton was directing traffic at the scene of a motor vehicle crash. Heavy fog was present at the scene. Two flatbed tow trucks were at the scene loading the involved vehicles for removal from the scene. Only one lane of traffic was open.

As he performed his duties, Fire Police Captain Pinkerton was struck by a passing car. He was treated at the scene and then transported to the hospital.

On April 19, 2003, Fire Police Captain Pinkerton died as the result of complications of his injuries. The cause of death was listed as severe pneumonia as a complication of a pedestrian motor vehicle accident.

The driver of the car that hit Fire Police Captain Pinkerton was cited for careless driving.

Union Fire Company Web site - www.unionfireco251.org

February 26, 2003 – 2000hrs
Curtis Walters, Firefighter
Age 52, Career
Northwest Arkansas Regional Airport Fire Department, Arkansas

Firefighter Walters was on duty in the fire station. As was his custom, Firefighter Walters went into the apparatus bay at approximately 2000hrs. After his Captain did not hear from him for approximately 1 hour, the Captain went to check on him.

Firefighter Walters was found unconscious, not breathing, and without a pulse. The Captain called for assistance on his radio and a number of nearby law enforcement, fire department, and rescue personnel responded. Despite their efforts, Firefighter Walters died.

The cause of death was a heart attack.

Firefighter Walters had been promoted to Captain but had yet to serve his first shift in that capacity.

March 8, 2003 – 1443hrs
Stephen Leigh McGregor, Firefighter
Age 62, Volunteer
Baird Volunteer Fire Department, Texas

Firefighter McGregor called his Fire Chief to request assistance with a grass fire that was burning out of control on his land. Firefighter McGregor had accidentally started the fire.

A brush truck from the Baird Volunteer Fire Department responded to the scene. Firefighters found a fire involving approximately 1 acre of land and 200 large round bales of hay. Firefighter McGregor was observed using his tractor to plow a firebreak ahead of the advancing flame front.

Firefighters in two brush trucks controlled the fires using handlines. Firefighters again observed Firefighter McGregor's tractor. They drove to the side of the tractor and saw that Firefighter McGregor seemed to be leaning over; the tractor was running but not moving.

Firefighters mounted the tractor and discovered that Firefighter McGregor was unresponsive. Firefighter McGregor was removed from the cab of the tractor to the ground, an ambulance was called, and CPR was started. Firefighter McGregor was connected to an AED and a shock was delivered. ALS care was initiated upon the arrival of the ambulance and continued during the transport to the hospital. Firefighter McGregor never responded to these measures and was pronounced dead at the hospital.

The cause of death was listed as sudden cardiac death.

The Texas State Fire Marshal has prepared a thorough report on this incident. The report is available at www.tdi.state.tx.us/fire/fmloddinvesti.html

---

March 9, 2003 – 2100hrs
Bruce H. Young, Firefighter
Age 60, Volunteer
Middlebury Fire Department, Vermont

Firefighter Young responded to his fire station in order to respond to an incident at a local college. He started the engine, suffered a massive heart attack, and slumped over the wheel.

Two other firefighters on the apparatus immediately went to the aid of Firefighter Young. He was transported by ambulance to the hospital where he was pronounced dead.

March 15, 2003 – 0700hrs
Faron H. O'Quin, Deputy Chief
Age 53, Volunteer
Lone Pine Fire Department, Louisiana

Deputy Chief O'Quin and members of his fire department responded to the scene of a crash involving a tanker truck carrying acid. They had been on scene all night awaiting a cleanup company. There was fog in the area, including around the incident scene. Traffic barriers were set up to divert traffic from the scene.

A vehicle that swerved to avoid the roadblock struck Deputy Chief O'Quin. The cause of death was listed as trauma.

March 15, 2003 – 1600hrs
Mike L. Stanley, Firefighter
Age 52, Volunteer
Salisbury Volunteer Fire Department, Missouri

Firefighter Stanley drove a fire department pumper to the scene of a wildland fire that was caused by wind-blown sparks from a friendly fire. When he arrived on-scene Firefighter Stanley prepared to pump water into another fire truck that was already on the scene. As the hose hookups were made, Firefighter Stanley collapsed.

CPR was started immediately and Firefighter Stanley was transported to the hospital where he was later pronounced dead. The cause of death was listed as a heart attack.

March 16, 2003 – 1100hrs
Gerald F. Schumann, Firefighter
Age 55, Volunteer
Madison Hose Company #1, Connecticut

Firefighter Schumann and other firefighters had just completed Sunday morning maintenance at their fire station. The firefighters had washed apparatus, cleaned the station, and washed floors.

After the work was complete, Firefighter Schumann and other firefighters gathered in the station's second floor to relax and talk. Suddenly, Firefighter Schumann collapsed. An ambulance was called and firefighters began to provide medical treatment for Firefighter Schumann.

Despite efforts at the fire station and in the ambulance, Firefighter Schumann was pronounced dead. The cause of death was listed as a heart attack.

March 18, 2003 – 0303hrs
Charles Lance Mathew, Lieutenant
Age 20, Volunteer
LaBelle-Fannett Volunteer Fire Department, Texas

Lieutenant Mathew responded to the scene of a motor vehicle crash in his personal vehicle. He was not wearing any reflective equipment or clothing.

Lieutenant Mathew arrived on the scene and parked his vehicle on the outside shoulder of the road opposite the scene of the crash. Lieutenant Mathew crossed the median and began to cross the road on the other side when a passing tractor-trailer truck struck him.

Lieutenant Mathew was thrown approximately 170 feet and came to rest in the median. He was severely injured and obviously deceased.

The Texas State Fire Marshal has prepared a thorough report on this incident. The report is available at www.tdi.state.tx.us/fire/fmloddinvesti.html

March 19, 2003 – 1940hrs
Thomas H. Kistler, Captain
Age 53, Volunteer
Polk County Fire District #1, Oregon

Captain Kistler was the front-seat passenger in an engine company. The engine was en route to a hazardous materials drill without siren or emergency lights. As a car approached from the opposite direction on a narrow road, the right rear wheels of the engine left the paved surface of the road and dropped into a ditch.

The driver was unable to steer the engine out of the ditch and traveled a total of approximately 165 feet before crashing into a tree. The apparatus struck the tree directly in front of Captain Kistler's seated position.

The driver and a passenger from the rear of the cab were unhurt and began to assess the injuries to Captain Kistler. He was found to be breathing but unresponsive and was trapped in the cab.

After approximately 1-1/2 hours, Captain Kistler was removed from the cab and pronounced dead at the scene. The extrication was made more difficult by the tree, equipment malfunctions, and the structure of the apparatus cab.

All three apparatus passengers were wearing their seatbelts. The cause of death for Captain Kistler was listed as positional asphyxia.

For additional information regarding this incident, please refer to NIOSH Fire Fighter Fatality Investigation and Prevention Program report F2003-14 (www.cdc.gov/niosh/face200314.html).

March 21, 2003 – 0850hrs
Oscar "Ozzie" Armstrong, III, Firefighter
Age 25, Career
Cincinnati Fire Department, Ohio

Firefighter Armstrong and the members of his fire company responded to the report of a fire in a two-story residence. The first fire department unit on the scene, a command officer, reported a working fire.

Firefighter Armstrong assisted with the deployment of a 350-foot, 1-3/4-inch handline to the front door of the residence. Once the door was forced open, firefighters advanced to the interior. The handline was dry as firefighters advanced; the hose had become tangled in a bush.

As the line was straightened and water began to flow to the nozzle, a flashover occurred. The firefighters on the handline left the building and were assisted by other firefighters on the front porch of the residence. All firefighters were ordered from the building, air horns were sounded to signal a move from offensive to defensive operations.

Several firefighters saw Firefighter Armstrong trapped in the interior by rapid fire progress. These firefighters advanced handlines to the interior and removed Firefighter Armstrong. A rapid intervention team assisted with the rescue.

Firefighter Armstrong was severely burned. He was transported by fire department ambulance to the hospital where he later died.

The Cincinnati Fire Department prepared a death investigation preliminary report related to this incident. The report is available at the fire department Web site for download.

The origin of the fire was determined to be a pan of oil on the stove.

Cincinnati Fire Department Web site - www.cincyfire.com

March 22, 2003 – 0500hrs
James Richard "Smitty" Smith, Firefighter
Age 62, Volunteer
Troy Volunteer Fire Department, Indiana

Firefighter Smith and the members of his fire department responded to an early morning structure fire involving a manufactured home. About 90 minutes into the incident, Firefighter Smith told another firefighter that he did not feel well. The other firefighter provided Firefighter Smith with some water and assisted him to an ambulance. Firefighter Smith told the ambulance attendants that he did not feel well and was having difficulty breathing and then suddenly collapsed. He was treated at the scene by paramedics and transported to the hospital.

The cause of death was listed as coronary artery thrombosis. The cause of the fire was listed as careless smoking.

March 25, 2003 – 2330hrs
Karlton Allen Cole Briscoe, Youth Firefighter
Age 16, Volunteer
Hickory Flat Volunteer Fire Department, Mississippi

Youth Firefighter Briscoe was responding to a motor vehicle crash in his personal vehicle. His vehicle left the roadway, crashed into a ravine, and rolled several times.

Firefighters responded to the scene from the original crash and extricated Youth Firefighter Briscoe. He was transported to the hospital where he died as a result of his injuries.

March 26, 2003 – 2120hrs
Kevin M. Whiteley, Training Officer
Age 46, Volunteer
Emmett City Fire Department, Idaho

Training Officer Whiteley and members of his fire department responded to a carbon dioxide incident at a residence. Training Officer Whiteley and another firefighter entered the residence with a carbon monoxide monitor; both firefighters were wearing full protective clothing and SCBA. Upon entry through two doors, they found high levels of carbon monoxide and left the structure.

Training Officer Whiteley assisted with raising a ladder to the roof of the residence and then removed his SCBA. Training Officer Whiteley changed his SCBA cylinder and began to complete an incident report.

Suddenly, Training Officer Whiteley dropped the clipboard that he had been holding and collapsed to the ground. Firefighters began to provide treatment, including CPR and oxygen. When an ambulance arrived, an AED was attached and delivered a total of six shocks. A normal heart rhythm was not restored and CPR was continued during the transport to the hospital.

Despite all efforts to revive him, Training Officer Whiteley was pronounced dead at the hospital. The cause of death was listed as complications of an enlarged heart. No significant carbon monoxide exposure was found.

March 27, 2003 – 1636hrs
Charles Guinn Krenek, Aviation Specialist
Age 48, Wildland Full-Time
Texas Forest Service, Lufkin, Texas

Aviation Specialist Krenek was operating his Texas Forest Service helicopter in support of the Space Shuttle Columbia recovery effort.

The helicopter was completing its second search mission of the day while hovering about 125 feet above the ground. During interviews with the NTSB investigator-in-charge, the surviving passengers reported that the helicopter lost power and descended rapidly into the 80-foot tall trees with no warning. The helicopter came to rest on its right side at the base of an 80-foot tree. The cockpit section of the fuselage was found crushed and the main cabin was mostly intact.

Aviation Specialist Krenek and the other crewmember were killed. Three passengers received serious but nonfatal injuries.

For additional information about this crash, consult the National Transportation Safety Board Web site at www.ntsb.gov/ntsb/query.asp - NTSB identification FTW03FA118.

March 30, 2003 – 0025hrs
Isaac Donald Tshudy, Fire Police Officer/Firefighter
Age 62, Volunteer
Lickdale Volunteer Fire Department, Pennsylvania

Fire Police Officer/Firefighter Tshudy responded to the scene of a crash involving a car and a tractor-trailer truck. Once he and other firefighters arrived on the scene, they quickly determined that there were no injuries as a result of the crash.

While other firefighters began to place road flares, Fire Police Officer/Firefighter Tshudy returned to the apparatus to start the generator. When firefighters did not hear the generator start, they investigated and found that Fire Police Officer/Firefighter Tshudy had collapsed.

Despite efforts to revive him, Fire Police Officer/Firefighter Tshudy died of a cardiac-related illness.

April 3, 2003 – 1302hrs
Richard A. Long, Firefighter
Age 32, Volunteer
Gallipolis Volunteer Fire Department, Ohio

Firefighter Long and the members of his department were dispatched to the report of a brush fire. Firefighter Long drove the first piece of fire apparatus to respond, a 2001 commercial chassis pumper/tanker with a 1,250-gallon water tank. He was followed by another firefighter in a brush truck. Firefighter Long was alone in the pumper/tanker.

As the pumper/tanker rounded a left-hand curve, the right wheels of the apparatus left the paved surface of the road. Firefighter Long steered to the left, the back end of the apparatus came around, the truck rolled over once, and came to rest on its right side.

Firefighters responding in the brush truck came upon the crash and rushed to the aid of Firefighter Long. He was trapped in the wreckage but was able to talk to the firefighters. They recited a prayer together and Firefighter Long lost consciousness and died.

Firefighter Long was pronounced dead at the scene. The cause of death was listed as compressive asphyxia.

Firefighter Long was not wearing his seatbelt. Unsafe speed was cited as a contributing factor in the crash.

For additional information regarding this incident, please refer to NIOSH Fire Fighter Fatality Investigation and Prevention Program report F2003-15 (www.cdc.gov/niosh/face200315.html).

April 7, 2003 – 2140hrs
Randy Hill, Battalion Chief
Age 43, Volunteer
Screven County Fire Department, Georgia

Battalion Chief Hill had just completed a 3-mile walk as a part of his monthly physical fitness training when he suddenly collapsed. Fellow firefighters came to his aid but were unable to revive him. The cause of death was listed as a heart attack.

April 8, 2003 – 0300hrs
Donald D. Maurice, Deputy Chief
Age 52, Volunteer
Wolcott Volunteer Fire Department, Company #1, Connecticut

Deputy Chief Maurice and other members of his department had just returned to the fire station after a response to a local convalescent home. Deputy Chief Maurice stowed his gear and collapsed of an apparent heart attack.

Other firefighters immediately began CPR and help was summoned. A police officer responded and used an AED but efforts to revive Deputy Chief Maurice were unsuccessful.

April 10, 2003 – 1815hrs
Vannie Duane Keever, Assistant Chief
Age 54, Volunteer
Gulfport/Gladstone Fire Protection District, Illinois

Assistant Chief Keever was the driver of a fire truck that responded to the scene of a structural fire in a vacant manufactured home. After arrival on the fire scene, Assistant Chief Keever began making connections to transfer water between apparatus.

The firefighter that was working with Assistant Chief Keever turned around and saw Assistant Chief Keever on the ground. Ambulance attendants on the scene came to his aid immediately. He was transported to the hospital where he was pronounced dead.

The cause of death was listed as occlusive coronary atherosclerosis.

April 13, 2003 – Time Unknown
Edward J. "Eddie" Weber, Fire Chief
Age 56, Volunteer
Elkhart Lake Fire Department, Wisconsin

Chief Weber had been fighting a series of wildland fires all day. After his department was dispatched to the third call of the day, Chief Weber arrived on the scene and began to coordinate an attack on a wildland fire.

Chief Weber suddenly collapsed. Firefighters from a neighboring department that were providing mutual aid began to provide treatment for Chief Weber. CPR was started immediately and an AED was used. Chief Weber was transported to the hospital; he could not be revived.

The cause of death was a heart attack.

April 14, 2003 – 2232hrs
John Robert "Bobby" Patrick, Lieutenant
Age 54, Career
Gwinnett County Department of Fire and Emergency Services, Georgia

Lieutenant Patrick and members of his ladder crew responded to the scene of a structure fire involving a butler-type building. The building housed a motorcycle repair shop.

After more than an hour on the scene, Lieutenant Patrick and his crew were directed to provide vertical ventilation. The crew went to the roof and prepared to open a hole. Lieutenant Patrick sat down on the roof and said that he was not feeling well. After a few moments, he collapsed.

He was removed from the roof by firefighters and brought to the ground. ALS-level care was provided immediately at the scene and in the ambulance while enroute to the hospital. Despite these efforts, Lieutenant Patrick was pronounced dead at the hospital.

Lieutenant Patrick's death was classified as a homicide; the fire was intentionally set.

The cause of death was listed as a cardiac dysrhythmia due to atherosclerotic coronary artery disease with superimposed physical exertion.

April 18, 2003 – 1030hrs
Mark Tyler Franklin, Engineer
Age 45, Career
Charlotte Fire Department, North Carolina

Engineer Franklin injured his knee while fighting a structure fire. He was off work and underwent surgery to correct the injury in October of 2003. During his recovery on November 30, 2003, Engineer Franklin died from a pulmonary embolism (blood clot) that resulted from the knee surgery. He was 46 at the time of his death.

The Charlotte Fire Department also suffered a firefighter fatality in January of 2003.

April 22, 2003 – 1100
Carl J. Mack, Firefighter and President
Age 68, Volunteer
New Chester Fire Department, Wisconsin

Firefighter Mack responded to the scene of a structure fire in a school that had been converted to apartments. He was assigned as the pump operator of one of the apparatus working at the scene.

Firefighter Mack collapsed of a heart attack. He was treated at the scene and transported to the hospital. He was pronounced dead at the hospital.

April 26, 2003 – 0844hrs
Bruce F. Spaulding, Firefighter/EMT-B
Age 49, Part-Time
Bourbonnais Fire Protection District, Illinois

Firefighter Spaulding was using a garden/lawn tractor to cut the lawn at the fire station. This task was one of the responsibilities of his position as a 12-hour Engineer.

At some point during the process, the tractor ran up against a short retaining wall. Firefighter Spaulding was thrown from the tractor and fell into a drainage ditch. As he reached the bottom of the ditch, Firefighter Spaulding suffered a severe head injury.

A firefighter arriving for duty discovered Firefighter Spaulding and called for help. Firefighter Spaulding was not breathing and had no pulse. ALS-level care was provided and he was rapidly transported to the hospital. Firefighter Spaulding was pronounced dead at the hospital due to his head injury.

April 28, 2003 – 1020hrs
Mark Austin Morgan, Ranger/Firefighter
Age 41, Wildland Full-Time
North Carolina Division of Forest Resources

Ranger Morgan was completing the annual "pack test" that is required for firefighters to participate in interagency wildland firefighting efforts.

Ranger Morgan complained of breathing problems during the test and stopped carrying the required 45-pound pack. He finished the 3-mile walk and did not show any outward signs of distress. Approximately 10 minutes after completing the test, Ranger Morgan suddenly collapsed.

Other rangers at the scene provided aid immediately and CPR was started after Ranger Morgan stopped breathing. An ambulance arrived approximately 20 minutes after the collapse and transported Ranger Morgan to the hospital. He was pronounced dead at the hospital due to a heart attack.

May 5, 2003 – 0830hrs
Harding O. Wentzell, Firefighter/Fire Police Officer
Age 81, Volunteer
Mexico Fire Department, Maine

Fire Police Officer Wentzell was responding in his personal vehicle to a structure fire. He suffered a heart attack and his car left the roadway. The car traveled for some distance before hitting a tree. The engine compartment of the car burst into flames.

As bystanders and a police officer pulled Fire Police Officer Wentzell from the car, an explosion occurred. Mutual-aid firefighters arrived after being diverted from their response to the original structure fire.

The firefighters extinguished the car fire and provided medical treatment for Fire Police Officer Wentzell. Paramedics arrived and continued to provide treatment. Fire Police Officer Wentzell was transported to the hospital and later pronounced dead.

May 10, 2003 – 2100hrs
Merlin J. Casey, Firefighter
Age 66, Volunteer
Mount Horeb Volunteer Fire Department, Wisconsin

Firefighter Casey was the driver on an ambulance that had responded to a medical incident. Firefighter Casey remained with the ambulance as other firefighters provided treatment inside a residence.

When other firefighters returned to the ambulance with the patient, they found Firefighter Casey slumped over the wheel. Firefighter Casey was removed from the ambulance and CPR was initiated.

Firefighter Casey was transported to the hospital but could not be revived. The cause of death was listed as a heart attack.

May 14, 2003 – 1430hrs
Richard G. Lupe, Lead Forestry Technician
Age 42, Wildland Full-Time
Bureau of Indian Affairs/Fort Apache Agency/Forestry/Fire Management, Safford, Arizona

Lead Forestry Technician Lupe was assisting with a prescribed burn involving approximately 2,000 acres on land that is part of an Apache Tribe reservation near White River. Lead Forestry Technician Lupe joined the effort during the third day of the burn.

An assigned lookout reported an increase in fire activity on the Southeast portion of the burn. Lead Forestry Technician Lupe advised that he would conduct reconnaissance of the area. A short time later, the area being investigated by Lead Forestry Technician Lupe experienced extreme fire behavior, evidenced by a large column of smoke.

Lupe transmitted calls to other firefighters advising them to back out of the area and to take refuge in safe areas that had already burned. The fire advanced quickly and overcame Lead Forestry Technician Lupe. He was unable to deploy his fire shelter due to the fact that embers on the ground would have burned him.

Nearby crews received radio calls from Lead Forestry Technician Lupe requesting medical assistance. He reported that he had been burned. Lupe was able to walk out of the area and meet firefighters.

Lead Forestry Technician Lupe was transported by firefighters to a helicopter landing area. He was flown to a local hospital and then flown again to a regional burn center. Lead Forestry Technician Lupe died on June 19, 2003, as a result of complications of his burns.

May 18, 2003 – 1130hrs
Jason Lee Ellis, Firefighter
Age 29, Paid-on-Call
Loretto Fire Department, Tennessee

Firefighter Ellis had been participating in a weekend training exercise at the Tennessee Fire Service and Codes Enforcement Academy. After the training activities were concluded, Firefighter Ellis was riding on the tailgate of a pickup owned and operated by another firefighter. The pickup was on the grounds of the Academy, traveling from the training site to the front of the campus. As the pickup accelerated, Firefighter Ellis fell from the vehicle and sustained a severe head injury.

Firefighter Ellis was treated at the scene and flown by medical helicopter to a regional hospital. Firefighter Ellis died on May 24, 2003, as a result of his head injuries.

May 22, 2003 – 2215hrs
Anndee Mikole Huber, Fire Explorer
Age 16, Volunteer
Newcastle Volunteer Fire Department, Wyoming

Fire Explorer Huber was a passenger in a tanker responding to a rekindle of some railroad ties that had burned earlier in the day.

As the tanker responded on a gravel road, the right wheels of the apparatus left the roadway, the driver over-corrected to the left, and the tanker rolled over.

Fire Explorer Huber was partially ejected in the rollover and pinned between the ground and the passenger's door. Responding firefighters used extrication equipment to provide access to Fire Explorer Huber and remove her from under the vehicle. Fire Explorer Huber was transported to the hospital and pronounced dead shortly after her arrival.

Fire Explorer Huber was not wearing a seatbelt at the time of the crash. Fire Explorer Huber drowned in water from the overturned tanker that collected in the ditch as the apparatus laid on its side.

The driver of the apparatus was under the influence of alcohol at the time of the crash. He had a history of alcohol abuse and had been seen in a local bar 15 minutes prior to the response. His blood alcohol level after the crash was found to be .16 percent, twice the legal limit. The driver was convicted of DUI and sentenced to 14 to 18 years in prison.

May 26, 2003 – 1725hrs
David L. Carbonneau, Sr., Fire Police Lieutenant
Age 49, Volunteer
Maytown-East Donegal Township Fire Department, Pennsylvania

Fire Police Lieutenant Carbonneau was driving the department's fire police unit between traffic points during the community's Memorial Day parade when he suffered a heart attack.

Fire department members rushed to his side and started CPR. An AED also was used in efforts to revive Fire Police Lieutenant Carbonneau. He was transported to a local hospital where later he was pronounced dead.

Maytown-East Donegal Township Fire Department Web site - www.maytownedfd.com

June 14, 2003 – 1929hrs
Donald Schreckengast, Fire Police Lieutenant
Age 55, Volunteer
Logan Fire Department, Pennsylvania

Fire Police Lieutenant Schreckengast had directed traffic during a community parade for over 1-1/2 hours. He drove the fire police vehicle back to the fire station. As the vehicle pulled into the station, Fire Police Lieutenant Schreckengast suffered a heart attack. He was transported to the hospital but was pronounced dead 35 minutes after he became ill.

Logan Fire Company Web site - www.loganfire.com

June 15, 2003 – 2100hrs
Trent Anthony Kirk, Lieutenant
Age 39, Career
Memphis Fire Department, Tennessee

Charles A. Zachary, Private
Age 39, Career
Memphis Fire Department, Tennessee

Lieutenant Kirk and Private Zachary were assigned to Engine Company 31. Lieutenant Kirk was working an overtime shift.

At 1946hrs, E31 and other Memphis Fire Department companies were dispatched to the scene of a structure fire involving a Family Dollar Store. As they arrived on-scene, they found smoke showing from the store at the end of a strip mall.

Lieutenant Kirk and a Lieutenant from another fire company proceeded through the retail area of the store and encountered only light smoke. When they attempted to enter a small office in the stock area at the rear of the store, they encountered a working fire. They were unable to close the office door and the fire advanced rapidly.

Private Zachary and other firefighters advanced handlines into the interior of the store and began fire suppression operations. As they worked in the rear of the structure, conditions worsened rapidly as dense smoke and high heat levels filled the building. Private Zachary requested relief and left the nozzle to return to the exterior. It is likely that he became disoriented in the smoke, although his actions after leaving the nozzle are unknown.

Lieutenant Kirk and another firefighter began to direct their hose stream into the stockroom area. They heard a firefighter call for help. A structural collapse occurred and knocked Lieutenant Kirk and the other firefighter to their knees. Lieutenant Kirk transmitted a Mayday call and said that he was trapped in the building. The collapse occurred approximately 17 minutes after the initial dispatch.

The firefighter with Lieutenant Kirk was able to free him from the debris, and both firefighters headed for the front of the store following their hoseline. As the firefighter crawled over a large pile of debris, he and Lieutenant Kirk lost contact.

Previous to the collapse, a rescue company had been assigned Rapid Intervention Crew (RIC) duties. Upon hearing Lieutenant Kirk's Mayday, the RIC advanced into the interior of the store and began their search. The RIC located and removed a firefighter; he was out of air and disoriented. The RIC then located the firefighter who had been with Lieutenant Kirk; he too was out of air and disoriented.

A ladder company was the only fire company at the rear of the building. They had forced entry to a rear door but did not have a handline and could not advance into the building. These firefighters heard an activated PASS device in the interior. After hearing reports of missing firefighters, the rear sector commander allowed firefighters to enter the interior without a handline to search for the downed firefighters. Upon entering the structure, firefighters heard two PASS devices. They were able to follow the sound to Private Zachary and remove him from the building. Upon his removal, ALS-level EMS procedures were initiated and he was transported to the hospital.

Firefighters made repeated rescue efforts but were driven from the store by rapid fire progress and their efforts were slowed by the structural collapse. Due to fire conditions, the IC ordered an end to all interior operations.

After the major body of fire was controlled with exterior streams, a rescue company breached a wall at the rear of the structure. The location of the hole was based on reports of the whereabouts of Lieutenant Kirk. He was removed from the building and transported to the hospital, where he was pronounced dead. Lieutenant Kirk received burns over 97 percent of his body; his carboxyhemoglobin level was 29 percent. The cause of death was listed as burns.

Private Zachary suffered severe surface and inhalation burns. Private Zachary died as the result of his thermal inhalation injuries on June 16, 2003.

The cause of the fire was determined to be arson. The store manager ignited the fire in an office to the rear of the structure. The fire was set to cover the theft of several thousand dollars from the store safe.

June 16, 2003 – 1830hrs
Randy Gene Utley, Firefighter/EMT
Age 30, Volunteer
Grayson Fire Department, Kentucky

Firefighter Utley and members of his fire department were working numerous emergency calls associated with storms that were passing through the area. Firefighter Utley left the scene of a motor vehicle collision to respond to a mutual-aid request for assistance in a flooded area with trapped civilians.

Firefighter Utley was responding along a two-lane road with his emergency equipment in operation. He was driving his personal pickup truck.

Firefighter Utley's pickup hydroplaned on the wet road surface, crossed over the opposing lane of the road, struck a ditch which turned the truck sideways, and collided with the support post of a billboard. The door on the driver's side of the pickup sustained the major impact of the crash.

A motorist that came upon the crash site called 911 and firefighters from the Grayson Fire Department responded to the scene. Firefighter Utley was extricated and transported to a trauma center. He was pronounced dead at the hospital.

Firefighter Utley was not wearing a seatbelt. Due to the lack of skid marks, his speed could not be estimated. The cause of death was listed as massive head and chest trauma.

For additional information regarding this incident, please refer to NIOSH Fire Fighter Fatality Investigation and Prevention Program report F2003-19 (www.cdc.gov/niosh/face200319.html).

June 26, 2003 – 1945hrs
Ralph Dwayne Dawdy, Assistant Chief/EMT/SAR Coordinator
Age 46, Volunteer
Animas Volunteer Fire and Rescue, New Mexico

Assistant Chief Dawdy was the driver of a military surplus 2-1/2 ton, 6 by 6 vehicle that had been converted to a fire department tanker. The water tank capacity was 1,200 gallons.

State-wide mutual-aid assistance had been requested for a wildland fire in a national forest. Assistant Chief Dawdy drove the tanker; he was followed by another firefighter in a fire department SUV. The pair planned to spend a week on the fire shuttling water to brush trucks on the fireline.

After traveling approximately 135 miles, Assistant Chief Dawdy was operating the tanker on a rough road within the forest. The road curved and sloped downhill. Assistant Chief Dawdy lost control of the tanker, it rolled over in the roadway, left the roadway and rolled over three more times before coming to rest on its wheels approximately 123 feet from the roadway in a canyon. Assistant Chief Dawdy was ejected during one of the early rolls and was crushed as the tanker rolled over him.

The other firefighter following in the SUV came upon the crash and searched for Assistant Chief Dawdy. When he located him, the firefighter determined that Assistant Chief Dawdy was dead.

Inspections of the tanker after the crash revealed that the brakes were 100-percent ineffective due to leaks at the master cylinder and slave cylinder. The brake fluid reservoir was empty at the time of the crash and a number of brake fluid containers were found aboard the tanker. In addition, the tanker was not equipped with seatbelts.

For additional information regarding this incident, please refer to NIOSH Fire Fighter Fatality Investigation and Prevention Program report F2003-23 (www.cdc.gov/niosh/face200323.html).

July 21, 2003 – 2105hrs
Samuel Lee Green, Firefighter
Age 50, Career
Shelby County Fire Department, Tennessee

Firefighter Green suffered a heart attack while on duty in the fire station. Firefighter Green was discovered unconscious by other firefighters after he had stepped outside of the fire station briefly.

Shelby County Fire Department Web site - www.shelbycountytn.gov

---

July 22, 2003 – 1525hrs
Jeffrey Clayton Allen, Firefighter
Age 24, Wildland Part-Time
United States Forest Service, Idaho

Shane William Heath, Firefighter
Age 22, Wildland Part-Time
United States Forest Service, Idaho

Firefighters Allen and Heath rappelled off of a helicopter into a rugged part of the Salmon-Challis National Forest. The two were charged with establishing a helicopter landing zone to facilitate efforts to fight the Cramer fire. The fire was started by a lightning strike.

The firefighters were dropped at the site at 0943hrs. They were contacted throughout the day by radio to assess their progress. At approximately 1500hrs, they made radio contact and requested that they be picked up. No helicopters were available at that time. A more urgent request for pickup was received at 1509hrs. At 1513hrs, the firefighters reported fire below them and that the fire was advancing toward them. A helicopter arrived at approximately 1524hrs but was unable to land due to smoke conditions.

Observers on other aircraft witnessed fire progression and extremely high flame fronts in the area of Firefighters Allen and Heath. Further attempts to contact the firefighters by radio were unsuccessful. The bodies of both firefighters were found together by other firefighters later in the day.

Both firefighters carried fire shelters but neither shelter was found fully deployed. The cause of death for both firefighters was listed as massive burns.

A full report on the incident prepared by the United States Forest Service is available at: www.fs.fed.us/fire/safety/investigations/cramer/report/fatual_rpt.pdf

July 25, 2003 – 1703hrs
Randall L. Harmon, Pilot
Age 44, Wildland Contract
Superior Helicopter, LLC, Oregon, under contract to the Bureau of Indian Affairs

Pilot Harmon was operating a twin-engine Kaman K-MAX helicopter while fighting the 2,200-acre Mc-Ginnis Flats fire near Keller, Washington, on the Colville Indian Reservation. The helicopter was carrying a 660-gallon bambi bucket.

The aircraft had recently been fueled and Pilot Harmon was flying his fourth mission of the day. Observers on the ground reported a change in the sound of the aircraft and Pilot Harmon radioed that he was in trouble.

The helicopter crashed and a subsequent fire consumed most of the wreckage. Pilot Harmon was killed upon impact.

For additional information about this crash, consult the National Transportation Safety Board Web site at www.ntsb.gov/ntsb/query.asp - NTSB identification SEA03GA153.

July 26, 2003 – 1035hrs
Randall Bonito, Jr., Firefighter/Helitack Crewmember
Age 32, Wildland Part-Time
Bureau of Indian Affairs Fire Management, Arizona

Jess Pearce, Pilot
Age 50, Wildland Contract
Airwest Helicopters under contract to the Bureau of Indian Affairs Fire Management, Arizona

Pilot Pearce was operating a Bell 206L-3 helicopter and flying a helitack crew to fight the Wilderness fire near Whiteriver, Arizona. There was a total of five people aboard the helicopter.

Two firefighters were dropped off in a meadow and the helicopter moved to another landing area to drop off an additional firefighter and some tools. As the third firefighter dismounted the helicopter, he heard a knocking noise. As the helicopter returned to the meadow dropoff point, it was observed to be flying slowly and at a low altitude. The helicopter passed the dropoff point and crashed into a wooded area.

Firefighters ran to the crash site and quickly extinguished a fire in the helicopter's engine area. Firefighter Bonito and Pilot Pearce were killed upon impact.

For additional information about this crash, consult the National Transportation Safety Board Web site at www.ntsb.gov/ntsb/query.asp - NTSB identification LAX03GA244.

July 28, 2003 – 1730hrs
Randy Neal Jones, Lieutenant
Age 23, Volunteer
Cool Springs Volunteer Fire Department, North Carolina

Lieutenant Jones and another firefighter were responding in Lieutenant Jones' personal vehicle, a 2000 Chevrolet pickup, to a report of a structure fire.

As they responded, the right wheels of the pickup left the roadway. Lieutenant Jones overcorrected to the left and the pickup began to slide sideways and rolled three times. It left the right side of the roadway and struck a power pole while airborne on the fourth roll. The vehicle then overturned a fifth time and came to rest on a side road. Both Lieutenant Jones and the other firefighter were ejected.

Firefighters responding to the structure fire were advised of the crash and some units were diverted to the crash scene. Firefighters found both Lieutenant Jones and the other firefighter injured. Lieutenant Jones was airlifted by helicopter from the scene but expired prior to his arrival at the hospital. The other firefighter was transported by ground ambulance and survived his injuries.

The police report on the crash estimated the pickup's speed prior to the crash at 80 miles per hour in a 55 mile per hour zone. Neither Lieutenant Jones or his passenger was wearing seatbelts. The cause of death for Lieutenant Jones was an acute intracranial (head) injury.

The fire that caused the initial response was found to have multiple points of origin.

Cool Springs Volunteer Fire Department Web site www.coolspringsvfd.org

The Cool Springs Volunteer Fire Department suffered a second firefighter fatality in September of 2003.

August 6, 2003 – 1216hrs
Jeffery Alan Koval, Firefighter
Age 41, Volunteer
Inkom Fire Department, Idaho

Firefighter Koval had been working a wildland fire with members of his department. He was detailed to go back into town and retrieve some supplies.

As he drove his city pickup back to the fire scene, he was involved in a single-vehicle crash. His car left the interstate highway; he crashed through a fence and then over an embankment. Firefighter Koval had to be extricated from the vehicle. He was transported to the hospital and pronounced dead.

Firefighter Koval was wearing his seatbelt.

August 6, 2003 – 1600hrs
Ronald L. Holmes, Firefighter/EMT
Age 43, Career
North Platte Fire Department, Nebraska

Firefighter Holmes was the driver of an ambulance during a long-distance transport of a patient to the hospital. As they traveled along an interstate highway, they came to a section of the highway where the left-hand lane was closed due to road construction.

The ambulance was behind a tractor-trailer truck that was proceeding slowly. The ambulance was struck from behind by another tractor-trailer truck and pushed into the rear of the truck in front of the ambulance.

Firefighter Holmes was crushed in the impact and was pronounced dead at the scene. The firefighter in the rear of the ambulance received severe injuries and was transported to the hospital. The patient was injured but the injuries were not life threatening.

North Platte Fire Department Web site - www.ci.north-platte.ne.us/fire

---

August 8, 2003 – 1000hrs
Wayne E. Mitchell, Firefighter Recruit
Age 37, Career
Miami-Dade Fire Rescue, Florida

Firefighter Recruit Mitchell and the members of his class were participating in a training session in a simulator designed to resemble a small ship. The class was in the 14th week of a 16-week program. Five recruits and three training instructors were in the interior of the structure wearing full structural protective clothing and SCBA.

As the exercise progressed, the heat in the interior of the simulator became intense and the instructors and several firefighters went to the exterior. Once outside, it was discovered that Firefighter Recruit Mitchell was still inside. The training officers reentered the simulator, located Mitchell, and removed him to the exterior.

When he was removed, Firefighter Recruit Mitchell was in cardiac and respiratory arrest. Firefighters from a fire station located at the training site began ALS-level emergency medical procedures and Firefighter Recruit Mitchell was transported to the hospital.

No official cause of death has been released. According to the medical examiner, Firefighter Recruit Mitchell had two minor heart conditions that were undetectable when he was alive but may have made him more susceptible to develop an abnormal heartbeat and contributed to his death. He was also found to have suffered severe burns to his hands and knees.

Four other recruits were transported to the hospital. They were treated for burns and heat exhaustion and released.

August 13, 2003 – 1350hrs
Barry D. Lutsy, Firefighter
Age 45, Volunteer
Racine Volunteer Fire Department, West Virginia

Firefighter Lutsy and other members of his department responded to a truck fire in their community. Once the fire was extinguished, firefighters returned to the fire station.

Firefighter Lutsy and another firefighter were draining and rolling hose at the front of the fire station. The driver of an engine apparatus pulled the apparatus forward to better align it to be backed into the apparatus bay. The apparatus was placed in reverse gear and began to back towards the station. According to the law enforcement report on the incident, Firefighter Lutsy and the other firefighter were talking but took notice of the backup alarm and returned to their task.

The apparatus continued in reverse for approximately 100 feet and came upon the firefighters, who had their backs toward the moving truck. The firefighters turned and saw the approaching truck just before impact. The firefighter working with Firefighter Lutsy was able to grab hold of the rear step as he fell and held onto it as the truck continued to move. The right rear tires of the truck rolled over Firefighter Lutsy and inflicted severe injuries.

Other firefighters immediately began to provide medical assistance to Firefighter Lutsy and the other injured firefighter. Both firefighters were transported to the hospital by ambulance. The injured firefighter was treated and released. Firefighter Lutsy was pronounced dead at the hospital.

Racine Volunteer Fire Department Web site - www.racinevfd.com

---

August 15, 2003 – 0200hrs
Wayne W. Mickle, State Firefighter I
Age 48, Wildland Full-Time
Massachusetts Wildfire Crew, Bureau of Forest Fire Control, Massachusetts Department of Conservation and Recreation

Firefighter Mickle and other members of his wildland firefighting crew were assigned to the Bole Meadow fire near Missoula, Montana. The crew had been working the fire for 4 days.

After a day of working the fire, he spoke with his wife by phone and complained that he was not feeling well. He told her that he was thinking of seeking medical attention but apparently decided to turn in for the night. He was found dead by other firefighters the next morning.

The cause of death was listed as a heart attack.

August 19, 2003 – 1912hrs
Edward Giacinto Buti, Inmate Firefighter
Age 54, Wildland Part-Time
Idaho Correctional Institute – Orofino

Inmate Firefighter Buti and the members of his crew were hiking up to the site of a wildland fire. The crew was assigned to the Sims fire near Elk City. The weather was extremely hot with temperatures in the low 100's.

As the team hiked, Inmate Firefighter Buti suddenly collapsed. Other firefighters performed CPR until the arrival of a medical team from a nearby forest service base. Despite their efforts, Inmate Firefighter Buti could not be revived.

The cause of death was listed as a heart attack.

August 24, 2003 – 1045hrs

Paul Eugene Gibson, Firefighter
Age 24, Wildland Contract
First Strike Environmental, Oregon

Richard Burt Moore, II, Firefighter
Age 21, Wildland Contract
First Strike Environmental, Oregon

David Kelly Hammer, Firefighter
Age 38, Wildland Contract
First Strike Environmental, Oregon

Leland David Price, Jr., Firefighter
Age 26, Wildland Contract
First Strike Environmental, Oregon

Jeffrey David Hengel, Firefighter
Age 20, Wildland Contract
First Strike Environmental, Oregon

Mark David Ransdell, Firefighter
Age 23, Wildland Contract
First Strike Environmental, Oregon

Jessie Dean James, Firefighter
Age 22, Wildland Contract
First Strike Environmental, Oregon

Ricardo Martin Ruiz, Firefighter
Age 19, Wildland Contract
First Strike Environmental, Oregon

These eight firefighters were a part of a 20-member contract firefighting crew. The firefighters were returning from 11 days of fighting fires in Idaho's Boise National Forest. The firefighters rode in a 16-passenger van as a part of a caravan of two vans and a truck.

As the van traveled through a two-lane sloped curve marked as a no passing area, it crossed the centerline to pass a tractor-trailer truck. The van struck a tractor-trailer truck traveling in the opposite direction head-on. When the vehicles came to rest, a fire began which consumed the van and the majority of the tractor-trailer. The occupants of the tractor-trailer truck were able to escape the fire but all eight firefighters remained in the burning wreckage of the van.

The crew boss was in the lead vehicle some distance ahead and did not witness the crash. When the crew boss lost radio contact with the firefighters, his vehicle pulled over to wait for them to catch up. Seeing smoke in the distance, they drove to the crash scene and found the van totally engulfed in fire.

As they traveled home, members of the crew had purchased beer and other food items in Idaho and again at a store just previous to the crash. Although tests were unable to determine exactly the amount of alcohol in the driver's blood, it is accepted that the driver of the van had been consuming alcohol before the crash.

All eight firefighters were either killed in the crash or in the ensuing fire and were pronounced dead at the scene.

First Strike Environmental was charged with 18 counts of reckless endangerment and drunken driving under Oregon's corporate responsibility law. The charges were dropped in February of 2004.

First Strike Environmental Web site - www.dwp.bigplanet.com/firststrike/firelineservices

August 27, 2003 – 1600hrs
Stephen G. Gavin, Lieutenant
Age 52, Volunteer
Owego Fire Department, New York

Lieutenant Gavin and the members of his fire department were fighting a fire in a local lumberyard. Lieutenant Gavin was in the process of doing overhaul after spending the preceding hour engaged in direct fire suppression.

The weather was hot and humid. Lieutenant Gavin complained of exhaustion and chest pain. He was treated by EMS workers on the scene and transported to the hospital. He suffered a heart attack and survived until September 2, 2003, when he died of complications of the heart attack.

September 6, 2003 – 1305hrs
Jerry Wayne Armstrong, Captain
Age 45, Volunteer
Bryant Volunteer Fire Department, Indiana

Captain Armstrong was working on a fire department fundraiser at the fire station. He told other workers that he was tired and went to sit under a tree, where he suffered a heart attack.

Firefighters came to his aid and provided assistance until an ambulance arrived. Captain Armstrong was transported to the hospital, where he was pronounced dead.

September 8, 2003 – 2253hrs
William F. Ramsey, Firefighter
Age 42, Volunteer
Connellsville Township Volunteer Fire Department, Pennsylvania

Firefighter Ramsey and other members of his department responded to a report of an unknown fire in their community. Upon their arrival on scene, firefighters found and extinguished a trash fire.

Firefighter Ramsey returned to the station and was beginning to remove his equipment when he collapsed of an apparent heart attack. Firefighters immediately went to his aid, and an AED was used in an attempt to revive him. Firefighter Ramsey was flown to a regional care center where he died about 3 hours after becoming ill.

The cause of death was listed as ventricular tachycardia due to severe cardiomyopathy. According to the American Heart Association, cardiomyopathy is a serious disease in which the heart muscle becomes inflamed and doesn't work as well as it should. There may be multiple causes, including viral infections.

September 20, 2003 – 0120hrs
William A. Wheeler, Fire Police Officer
Age 66, Volunteer
Saxton Volunteer Fire Department, Pennsylvania

Fire Police Officer Wheeler was responding to a motor vehicle crash with other members of his fire department. He was driving his personal vehicle, a pickup truck. As he responded to the scene, Fire Police Officer Wheeler suffered a heart attack. His vehicle left the roadway and crashed into a utility pole. The airbag deployed during the crash but there was no damage to the interior of the cab.

Firefighters working at the scene of the original crash were informed of another crash involving a fire department member. Firefighters responded to the scene and found Fire Police Officer Wheeler behind the wheel in respiratory and cardiac arrest. Fire Police Officer Wheeler was rapidly extricated and provided with ALS-level emergency care on the scene. He was transported to the hospital where he was later pronounced dead.

The cause of death was listed as a cardiac arrhythmia with congestive heart failure due to a recent myocardial infarction due to occlusion of a coronary artery.

Fire Police Officer Wheeler was wearing his seatbelt at the time of the crash.

September 27, 2003 – Time Unknown
Jon Bill Luttman, Fire Chief
Age 53, Volunteer
Redkey Volunteer Fire Department, Indiana

Chief Luttman had worked at a fire department fundraiser on the morning of September 27th. He complained of not feeling well at the event and went home to rest. He drove himself to the hospital complaining of flu-like symptoms. He suffered a heart attack and died that day.

September 27, 2003 – 1745hrs
James J. O'Shea, Firefighter 1st Grade
Age 40, Career
Fire Department City of New York, New York

Firefighter O'Shea responded as a member of a ladder company to a fire that involved stacked cardboard boxes outside of an apartment complex. Firefighter O'Shea assisted with extensive overhaul and complained of not feeling well at the scene.

He declined medical assistance and went off-duty shortly after returning to his firehouse. He drove himself home and was found unconscious in the driveway, the victim of a heart attack. His family rushed him to the hospital but he could not be resuscitated.

A neighborhood teenager intentionally started the fire. The teen was arrested and charged with arson.

The Fire Department City of New York suffered a second firefighter fatality in December of 2003.

September 30, 2003 – Time Unknown
Gerald Hayswood Williams, Firefighter
Age 60, Volunteer
Cool Springs Volunteer Fire Department, North Carolina

Firefighter Williams participated in an evening training session involving hose evolutions. He later went to bed in the fire station and was found dead by other firefighters the next morning, the victim of an apparent heart attack.

Cool Springs Volunteer Fire Department Web site www.coolspringsvfd.org

The Cool Springs Volunteer Fire Department suffered another firefighter fatality in July of 2003.

September 30, 2003 – 1510hrs
George D. Petrosky, Fire Police Officer
Age 78, Volunteer
Matawan Borough Fire Department, New Jersey

Fire Police Officer Petrosky was directing traffic at the scene of a motor vehicle crash. He suffered a heart arrhythmia and collapsed. When he fell, he struck his head on the pavement and suffered a serious head injury.

The following day, surgery was performed to stop bleeding in his brain. He lapsed into a coma and never regained consciousness. Fire Police Officer Petrosky died on November 9, 2003, as a result of his injuries.

October 1, 2003 – 0850hrs
John August Garman, Firefighter
Age 42, Volunteer
New Bremen/German Township Fire Department, Ohio

Kenneth Joseph Jutte, Firefighter
Age 44, Volunteer
New Bremen/German Township Fire Department, Ohio

Firefighters in New Knoxville responded to a report of a fire in an 80-foot tall concrete silo. The silo was 20 feet in diameter and contained sawdust and wood scrap from a manufacturing business. The sawdust and wood scrap were accumulated in the silo and then conveyed by mechanical equipment for use as fuel for an onsite electrical generation and heating plant. The manufacturing occupancy had been the site of numerous previous fires in all areas of the operation.

Firefighters observed smoke coming from a room at the base of the silo and from the top of the silo. They observed burning wood materials falling out of the bottom of the silo. They also used a thermal imaging camera and observed hot spots close to the base of the silo and about 15 feet above the base of the silo. Roof access was limited to a small ladder that scaled the side of the silo. The decision was made to request mutual aid from the New Bremen/German Township Fire Department for a ladder truck.

Prior to the arrival of the ladder truck, firefighters operated handlines into access doors at the base of the silo in an attempt to control the fire close to the bottom of the silo.

Firefighter Garman and Firefighter Jutte responded as a part of the crew on the New Bremen/German Township Fire Department tower ladder quint apparatus. Upon their arrival on the scene, they helped set up the ladder and bring a member of the New Knoxville Fire Department to the roof of the silo.

Once on the roof, firefighters observed smoke coming from an open roof hatch. The firefighters returned to the ground. The decision was made to attempt to extinguish the fire. A piercing nozzle was inserted through an access door near the ground. A handline with a cellar (distributing) nozzle from the aerial tower was operated into the roof hatch. A decision was made to switch from the cellar nozzle to a fog nozzle. The hose was shut down and removed from the hatch. Three firefighters were either on the roof of the silo or in the platform of the tower ladder.

As the uncharged hose was removed from the hatch (approximately 80 minutes after the ladder truck arrived on the scene) an explosion occurred. The force of the explosion ripped the roof of the silo off and damaged the top of the silo. Pieces of the silo rained down in an area as wide as 100 yards from the base of the silo. Firefighter Jutte and a New Knoxville firefighter on the roof of the silo were thrown to the ground and Firefighter Garman was blown off of the tower ladder platform and fell to the ground.

Firefighters on the scene began to provide medical assistance for the injured and EMS responders were called to the scene. Several of the victims were transported to the hospital by helicopter.

Firefighter Garman and Firefighter Jutte were killed. Ten others on the scene, including civilians, were injured.

Fire investigators concluded that the fire was caused by heat generated by the failure of mechanical equipment in the base of the silo.

October 3, 2003 – 1130hrs
Carl Dolbeare, Pilot
Age 54, Wildland Contract
Minden Air Corporation, Minden, Nevada – Under contract to the United States Forest Service

John Attardo, Co-Pilot
Age 51, Wildland Contract
Minden Air Corporation, Minden, Nevada – Under contract to the United States Forest Service

Pilot Dolbeare and Co-Pilot Attardo were the flight crew of a Lockheed P2V Neptune airtanker. The aircraft was being repositioned from Prescott, Arizona, to San Bernardino, California, after completing firefighting duty in Arizona.

Observers on the ground report seeing the plane flying at a low altitude through some clouds and the plane was seen making a steeper than normal 180-degree turn. The wings of the aircraft leveled and the plane was seen flying through a cloud; it briefly reappeared, and then entered the cloud layer. About 2 minutes after losing sight of the plane, smoke was observed. The crash was reported to local authorities.

About 2 hours after the crash, searchers located the wreckage. They found the wreckage and surrounding vegetation on fire. Initial responders reported that the area was cloudy and that visibility was limited. The crash occurred about 8 miles from the destination airport.

For additional information about this crash, consult the National Transportation Safety Board Web site at www.ntsb.gov/ntsb/query.asp - NTSB identification LAX04FA002.

Minden Air Corporation Web site - www.mindenair.com

October 7, 2003 – 0759hrs
James Phillip Allen, Firefighter
Age 43, Career
Philadelphia Fire Department, Pennsylvania

Firefighter Allen and the other members of his ladder company responded to a structure fire in a two-story residence. Upon their arrival, Firefighter Allen went to the roof of the structure by ground ladder for ventilation.

After removing the glass in some windows, Firefighter Allen suffered a heart attack. He pressed the distress button on his portable radio and collapsed. An upper-floor resident observed him from across the street. That resident also called 911 to report an injured firefighter.

Other firefighters came to the aid of Firefighter Allen and initiated CPR. Paramedics arrived and continued efforts as he was removed from the roof and transported to the hospital. Their efforts, however, were unsuccessful and Firefighter Allen was pronounced dead at the hospital.

The cause of death was listed as ischemic heart disease (narrowed heart arteries) with smoke inhalation listed as a significant condition.

A 7-year-old child playing with matches started the fire in a mattress.

October 14, 2003 – 1045hrs
Richard Warren "Dick" Black, Pilot
Age 57, Wildland Contract
Weyerhaeuser - Eugene Helicopter Operations, Oregon

David Craig Mackey, Forest Unit Supervisor
Age 53, Wildland Full-Time
Oregon Department of Forestry

Pilot Black and Forest Unit Supervisor Mackey were involved in a reconnaissance flight to locate water supplies for future fire suppression operations. They were flying in a Weyerhaeuser-operated Bell 206B helicopter. The pair located water supply locations that could be used by helicopters and recorded their GPS location for future use.

After about an hour in flight, the helicopter was proceeding along a river when it collided with a 5/8-inch neutral/ground cable line that was strung across the Siuslaw River. The weather at the time was cloudy and foggy.

The helicopter fell approximately 250 to 300 feet into the river and came to rest in about 3 feet of water. The cable was entangled in the wreckage. Pilot Black and Forest Unit Supervisor Mackey died from multiple traumatic injuries.

The cable that was struck by the helicopter was not marked as a flight hazard on maps. The same cable run was involved in a nonfatal helicopter crash in 1974. The cables have been in place since 1956 and have never been marked.

For additional information about this crash, consult the National Transportation Safety Board Web site at www.ntsb.gov/ntsb/query.asp - NTSB identification SEA04LA005.

October 20, 2003 – 1600hrs
James D. Richards, Assistant Fire Chief
Age 49, Volunteer
Oran Fire Department, Iowa

Assistant Chief Richards and members of his fire department were called to a fire involving harvested corn stalks that were ignited by sparks from a fire pit. The wind was blowing at 30 to 45 miles per hour.

Assistant Chief Richards was making connections to draw water from a portable water tank when he had a heart attack. Despite the efforts of firefighters on the scene, Assistant Chief Richards was not revived.

The cause of death was listed as a heart attack due to severe arteriosclerosis.

October 24, 2003 – 0846hrs
Ricardo A. "Ricky" Gonzales, District Chief
Age 47, Career
Beaumont Fire/Rescue, Texas

District Chief Gonzales responded to a crash involving a fire truck at 0846hrs. The fire truck overturned and three firefighters were injured. After transporting the firefighters to the hospital in his command vehicle, District Chief Gonzales responded to an automatic fire alarm at 1042hrs.

Later in the morning, District Chief Gonzales complained of pain in his neck, a headache, and flu-like symptoms. He did not feel that he could safely make the 60-mile commute to his home, so he slept at the fire station until approximately 1400hrs when he left on sick leave and went home. He felt so poorly when he arrived in his home town that he drove directly to his physician's office.

District Chief Gonzales was diagnosed as having suffered a heart attack. He was given medication and admitted to the hospital. He was treated in the hospital over 5 days and received an angioplasty and four stents. He was released from the hospital on October 29, 2003, and went home to recover.

On November 5, 2003, his wife found District Chief Gonzales dead at home. The cause of death was listed as a heart attack.

October 24, 2003 – 1530hrs
Darrell Keith Michael, President
Age 60, Volunteer
Russell Volunteer Fire Department, Pennsylvania

President Michael and members of his fire department were on stand-by at their fire station for a mutual-aid structure fire. The firefighters decided to prepare for a fundraiser to be held the next evening while they waited.

President Michael climbed a stepladder to get something off of a shelf. Other firefighters heard a loud crash and found President Michael unconscious on the floor.

Firefighters and EMS personnel provided care for President Michael and he was transported to the hospital. He was placed on life support and died as a result of his injuries on November 3, 2003.

Russell Volunteer Fire Department Web site - www.rvfd.home.westpa.net

October 27, 2003 – 2322hrs
Don Joseph Billig, Assistant Chief
Age 49, Volunteer
St. Cloud Fire Department, Minnesota

Assistant Chief Billig and members of the St. Cloud Fire Department Volunteer Division were dispatched to a report of a smoking generator at a road construction site. Firefighters discovered a problem with a dewatering unit being used at the site and contacted the contractor. Assistant Chief Billig and other firefighters stood by at the scene to await the arrival of the contractor.

The contractor arrived at the scene and firefighters prepared to return to quarters. Assistant Chief Billig and other firefighters were outside the vehicle, replacing traffic control barriers. A firefighter observed a vehicle headed toward the crew at a high rate of speed. He shouted for the other firefighters to look out for the approaching vehicle. Two firefighters were able to jump to safety but Assistant Chief Billig was struck when the vehicle failed to negotiate the construction detour.

Assistant Chief Billig was dragged under the vehicle, a full-size pickup, for 50 to 60 feet. Firefighters immediately rendered aid and requested assistance. Paramedics at the scene treated Assistant Chief Billig until it was obvious that he could not be revived.

The driver of the pickup fled the scene. He turned himself into police the next morning. The driver was charged with criminal vehicular homicide. Alcohol use and speeding likely were involved in the crash, according to law enforcement officials. The driver had two prior convictions for driving while intoxicated, and seven speeding tickets.

St. Cloud Fire Department Web site - www.ci.stcloud.mn.us/Web/departments/Fire/Fire.htm

October 28, 2003 — Time Unknown
Charles Tozzo, Captain
Age 55, Volunteer
Cold Spring Harbor Volunteer Fire Department, New York

Captain Tozzo was attending an officers' meeting in the fire station. During the meeting, he suffered a heart attack and later died.

October 29, 2003 — 1300hrs
Steven Liss Rucker, Engineer
Age 38, Career
Novato Fire Protection District, California

Engineer Rucker and the members of his engine company were assigned through a statewide mutual-aid system to the Cedar Fire near San Diego. They were assigned to defend a home in a hilly area that was threatened by wind-driven fire spread.

Upon their arrival at the home, the crew cleared some brush and decided that the location was defendable. The crew burned areas of brush near the house and made other preparations for the approach of the fire including laddering the house and stretching hoselines.

About 20 minutes after arriving at the house, the engine crew observed an increase in the fire activity below them. The fire made a half-mile run directly at Engineer Rucker and his crew in less than 2 minutes.

As the fire progressed toward the crew, the firefighters retreated to a defensive position behind the engine and operated 2 handlines for protection. Fire conditions worsened and the Captain of the crew observed flame lengths of 40 to 50 feet. Due to the intense heat, the Captain ordered his crew to abandon their position and seek shelter in the house. The firefighters, facing severe thermal exposure, ran for shelter to the rear of the house.

Two firefighters arrived safely to the interior of the house. They then heard a radio call indicating that a firefighter was down. The firefighters left the house and began to retrace their steps back toward the engine. They encountered the Captain, who told them that Engineer Rucker was down and needed their help. Due to intense heat, the firefighters and the Captain were unable to go to the aid of Engineer Rucker.

Despite efforts by all three members of the crew, Engineer Rucker could not be rescued. By the time firefighters were able to reach him, Engineer Rucker had expired. Apparently, Engineer Rucker tripped while making a run for the house. His Captain attempted to help him but was unsuccessful. Engineer Rucker died of burn injuries. The Captain received severe burn injuries and was hospitalized for an extended period of time.

The Cedar Fire eventually consumed 280,278 acres and destroyed 2,232 structures, 22 commercial buildings, and 566 outbuildings. Thirteen civilians were killed as a result of the fire and there were 107 injuries.

A link to the California Department of Forestry and Fire Protection "Green Sheet" report on the death of Engineer Rucker is posted on the Novato Fire Protection District Web site at www.novatofire.org/cedar_fire.aspx

Novato Fire Protection District Web site - www.novatofire.org

October 30, 2003 – 1543hrs
Roy Prouty, Fire Chief
Age 39, Volunteer
Country Lakes Volunteer Fire Company, Browns Mills, New Jersey

Chief Prouty and members of his fire department responded to a neighboring community to assist with emergency services at the scene of a motor vehicle crash.

Chief Prouty established a landing zone for medical helicopters. He helped load a patient into a helicopter, then suddenly collapsed of an apparent heart attack. He was treated at the scene and transported to the hospital. Despite all efforts, he was pronounced dead at the hospital.

Country Lakes Volunteer Fire Company Web site - www.countrylakesfire.org

November 8, 2003 – 1451hrs
Matthew Karl Brimer, Firefighter
Age 18, Volunteer
Weaver Fire Department, Alabama

Firefighter Brimer was responding in his personal vehicle to a report of a structure fire. As he drove around a curve, he lost control of the vehicle, skidded 226 feet, and sideways into an oak tree.

Firefighters responding to the initial incident were directed to the scene of the crash and were not aware that the incident involved Firefighter Brimer until they noticed a hat in the car. Firefighter Brimer was extricated by his fellow firefighters. He was pronounced dead at the scene.

Firefighter Brimer was wearing his seatbelt at the time of the crash. The law enforcement report on the incident cited an estimated speed of 70 miles per hour in a 40 mile-per-hour zone.

Firefighter Brimer had just joined the fire department the previous May upon his 18th birthday. The structure fire that caused the initial response was found to be minor.

November 17, 2003 – 1712
Richard James Tiffany, Paramedic Firefighter
Age 35, Career
Clark County Fire District #12, Washington

Paramedic Firefighter Tiffany was completing on-duty physical training at his fire station. He completed a treadmill run and collapsed of an apparent heart attack.

A company officer who also was engaged in physical training called 911 and began CPR. The company officer administered oxygen and applied an AED. Despite his efforts, Paramedic Firefighter Tiffany was pronounced dead at the fire station.

Clark County Fire District #12 Web site - www.ccfd12.org

November 17, 2003 – 2150hrs
Gary D. Boyert, Captain
Age 53, Career
Kansas City Fire Department, Kansas

Captain Boyert was responding in a fire department sedan to the report of a structure fire.

Roads were slick from recent rain. As his vehicle entered a curve, Captain Boyert lost control. The vehicle left the roadway and struck a utility pole and a fire hydrant. The vehicle came to rest on its roof. The driver's door of the car received the majority of the impact damage with the telephone pole.

Firefighters responded to the crash scene and provided medical care for Captain Boyert. He was extricated and then transported to the hospital, but could not be revived.

The preliminary cause of death was listed as positional asphyxiation. The alarm that generated Captain Boyert's response was a false alarm.

Kansas City Fire Department Web site - www.wycokck.org/departments/fire/index.html

November 19, 2003 – 0730hrs
Charles Forest Flowers, Sr., Assistant Fire Chief
Age 60, Volunteer
New Caney Volunteer Fire Department, Texas

Assistant Chief Flowers and the members of his department responded to a motor vehicle crash that required extrication. Assistant Chief Flowers was engaged in the extrication when he suddenly collapsed of a heart attack.

Assistant Chief Flowers was treated at the scene and transported to the hospital. He remained in the hospital until December 17, 2003, when he was discharged to continue his recovery at home. Assistant Chief Flowers died as the result of a heart attack that he suffered in his sleep on December 29, 2003.

November 24, 2003 – 1718hrs
Jeffery Allen Tiegs, Firefighter
Age 47, Volunteer
Amherst Fire District, Wisconsin

Firefighter Tiegs was the driver of an engine apparatus that responded to assist at the scene of a motor vehicle crash. As he dismounted the apparatus, Firefighter Tiegs was struck with a heart attack and collapsed.

Firefighters immediately began to provide assistance, including CPR and the use of an AED. An ambulance arrived and transported Firefighter Tiegs to the hospital. Firefighter Tiegs never regained consciousness and was pronounced dead at the hospital.

November 29, 2003 – 0530hrs
Martin H. McNamara, V, Firefighter
Age 31, Paid-on-Call
Lancaster Fire Department, Massachusetts

Firefighter McNamara and members of his fire department responded to the scene of a structure fire involving a 2-1/2-story wood balloon-frame residential building that contained multiple apartments. Two additional 1-1/2-story buildings were attached to the rear of the main building.

Firefighters found a working fire. Firefighter McNamara was assigned as a part of a crew that advanced an attack line into the basement of the structure. After a series of explosions, the firefighters were forced to leave the building.

Once outside, a head count was completed and Firefighter McNamara was discovered missing. Firefighters immediately reentered the basement; they could hear the chirp of Firefighter McNamara's PASS device but could not reach him due to fire conditions. After the fire was controlled, a rescue team entered the structure and located the body of Firefighter McNamara.

The cause of death was listed as smoke and soot inhalation. Firefighter McNamara also suffered facial burns prior to his death. Three other firefighters were injured in the fire; including a deputy chief who suffered severe smoke inhalation during an attempt to rescue Firefighter McNamara.

The cause of the fire was identified as the overheating of a power strip and extension cord in the basement.

November 30, 2003 – 1320hrs
Thomas W. DiOrio, Fire Police Lieutenant
Age 70, Volunteer
West Whiteland Fire Company, Pennsylvania

Fire Police Lieutenant DiOrio and members of his fire department responded to a structure fire in a townhouse. Fire Police Lieutenant DiOrio set up traffic barriers approximately one block from the incident scene. After he completed the setup of the barriers, he collapsed of a heart attack.

EMS and law enforcement personnel responded immediately to assist Fire Police Lieutenant DiOrio. Despite their efforts, he could not be revived.

December 1, 2003 – 2230 hrs
Nadar Hammett, Emergency Response Technician
Age 29, Career
Prince George's County Fire/EMS Department, Maryland

Emergency Response Technician (ERT) Hammett was driving home from department-mandated paramedic training. ERT Hammett's vehicle was struck by another vehicle, causing it to leave the roadway, roll over, and hit a tree.

The driver of the other vehicle left the scene but returned about 30 minutes later and reported to the police. ERT Hammett was wearing his seatbelt.

December 3, 2003 – 1208hrs
Charles Wayne Dillon, Captain
Age 40, Volunteer
Washington Parish Fire District #9, Louisiana

Captain Dillon was the driver and sole occupant of a 2,000-gallon water tanker. Captain Dillon was engaged in driver training duties.

As the tanker entered a left-hand curve, the right wheels of the apparatus left the paved surface of the road. Captain Dillon attempted to correct by steering to the left, the rear of the tanker came around, and the water tank separated from the apparatus. The cab of the apparatus began to roll and Captain Dillon was ejected.

The speed of the tanker was estimated at 55 miles per hour in a 55 mile-per-hour zone. A curve warning sign on the approach to the curve suggested a speed of 40 miles per hour.

An inspection of the tanker chassis after the crash revealed a number of mechanical problems with the truck. The right-front shock was not connected to the frame at its upper mount, the right rear brake was inoperative due to being out of adjustment, the right front brake was inoperative and contaminated with oil, and the left front brake was contaminated with oil.

As described in the law enforcement report on the crash, the tanker was most likely locally made or modified from some other use. Captain Dillon was ejected during the crash and received a fatal head injury.

December 3, 2003 – Time Unknown
Todd "Bubba" Dicks, Firefighter
Age 37, Volunteer
Warsaw Fire Department, Illinois

Firefighter Dicks was a passenger in a fire truck responding to the scene of a school bus/pickup truck crash in his community. During the response he appeared ill to other firefighters. Upon their arrival on the scene, EMS responders assigned to the original incident provided medical attention.

Firefighter Dicks was transported by ambulance to a local hospital. He was pronounced dead shortly after arrival. The cause of death was listed as a heart attack.

December 6, 2003 – 1616hrs
Ronald W. Fitzpatrick, Firefighter/Training Officer
Age 68, Volunteer
Long Branch Fire Company, New Jersey

Firefighter/Training Officer Fitzpatrick and members of his department were fighting a multiple-alarm fire in a vacant commercial building.

During the effort, Firefighter Fitzpatrick was in charge of the Rapid Intervention Team (RIT) that was in place at the fire. He collapsed from a heart attack. Efforts to revive him at the scene and in the hospital were unsuccessful.

The fire was reported to have been caused by arson.

Firefighter Fitzpatrick retired in 1986 as a Captain after 27 years with the Newark, New Jersey Fire Department.

December 13, 2003 – 1454hrs
George Oliver Wohl, Firefighter
Age 59, Volunteer
Congers Fire Department, New York

Firefighter Wohl was installing holiday lights on his residence when his fire department was dispatched to an emergency. Firefighter Wohl was on a ladder near the roofline of the home.

Firefighter Wohl slipped and fell to the ground as he began his response. His wife discovered he had fallen a few moments later and called 911. Firefighter Wohl was unconscious but breathing and he suffered a significant head wound.

Firefighter Wohl was transported to the hospital first by ambulance to a landing site and then by medical helicopter. He died at the hospital later that day. The cause of death was listed as fractures of the skull and ribs, brain hemorrhages, and contusions of the brain.

The source of the original dispatch was an automatic fire alarm. The alarm turned out to be false.

Congers Fire Department Web site - www.congersfd.org

December 16, 2003 – 1300hrs
Thomas Christopher Brick, Firefighter 4th Grade
Age 30, Career
Fire Department City of New York, New York

Firefighter Brick and the members of his ladder company were in a grocery store shopping for their meal when they were dispatched to a structure fire in a mattress warehouse.

Units arrived on the scene and found a working fire in a multistory brick and joist nonfireproof building. As the first-due ladder company, Firefighter Brick and the members of his crew used a saw to force entry to the building and then proceeded to the fire floor (second floor) for search and rescue.

Due to the volume of fire and greatly reduced visibility, the Incident Commander (IC) ordered all firefighters to leave the building. After he was outside of the building, Firefighter Brick's company officer found him to be missing. The IC was notified immediately. At the same time, another firefighter was located and believed to be Firefighter Brick. When the identity of this firefighter was confirmed, firefighters reentered the building to search for Firefighter Brick.

After several minutes of searching, Firefighter Brick was located. He was not breathing and had no pulse. CPR was initiated on the fire floor and continued to the hospital. Firefighter Brick was pronounced dead at the hospital.

The cause of death was listed as smoke inhalation. The concentration of carboxyhemoglobin in Firefighter Brick's blood at autopsy was 61 percent.

Firefighter Brick was a member of the first recruit class trained after September 11, 2001.

December 18, 2003 – 0815hrs
Dixie Lee Steckelberg, Lieutenant
Age 60, Volunteer
Lovilia Fire & Rescue, Iowa

Lieutenant Steckelberg was providing treatment on the scene of a motor vehicle crash. She helped move a patient onto a backboard and into an ambulance.

Lieutenant Steckelberg began to complain of shortness of breath and went to sit in the front-seat of the rescue vehicle. She was treated by EMT's and paramedics at the scene and again in the ambulance on the way to the hospital.

While enroute to the hospital in the ambulance, Lieutenant Steckelberg went into cardiac arrest. CPR was initiated and continued until she arrived at the emergency room. Despite additional efforts in the hospital, Lieutenant Steckelberg was pronounced dead at 0848hrs.

The cause of death was listed as a heart attack.

December 18, 2003 – 1730hrs
Thomas Frank Brown, Fire Specialist
Age 55, Career
Baltimore County Fire Department, Maryland

Fire Specialist Brown worked a 14-hour night shift at the Dundalk Fire Station. During the shift, he responded to several calls, including a working dwelling fire and a cardiac arrest, where he assisted with resuscitation on scene and enroute to the hospital. Fire Specialist Brown was relieved the morning of December 18th at 0630hrs. He was scheduled to return to work that same day by 1800hrs. When he failed to report to work, the engine company closest to his home was sent to investigate. Firefighters found him in bed, obviously deceased.

The cause of death was listed as a heart attack due to arteriosclerosis.

December 21, 2003 – 0758hrs
Benjamin Craig Rouchon, Firefighter
Age 25, Volunteer
Bluff Creek Fire Protection Territory, Louisiana

Firefighter Rouchon was at home when his fire department was paged to respond to assist a local ambulance service at an EMS incident. Firefighter Rouchon responded in his personal pickup truck.

During the response, Firefighter Rouchon lost control of his vehicle, crossed over the centerline, bounced off of a tree, crossed the roadway again, and struck an embankment.

Firefighters and ambulance personnel responded to the scene and provided treatment for Firefighter Rouchon. Despite their efforts, Firefighter Rouchon died of massive head injuries.

Firefighter Rouchon was not wearing his seatbelt at the time of the crash.

December 24, 2003 – 0530hrs
Shane Scott Brown, Firefighter/Paramedic
Age 25, Career
DeSoto Parish Fire District 8, Louisiana

Firefighter/Paramedic Brown was responding to an EMS incident in a fire department pickup truck.

As he responded through an unguarded railroad crossing, his vehicle was struck by a freight train. The vehicle was dragged approximately 60 feet and caught fire.

Just prior to impact, the train was estimated to be traveling at 40 miles per hour. Witnesses said that the train's warning lights and horns were in use at the time of the crash.

Firefighter Brown was wearing his seatbelt and he was partially ejected from the cab of the pickup. Firefighter Brown was pronounced dead at the scene of traumatic injuries.

Firefighter Brown is the second Fire District 8 firefighter to die in a train accident while on duty. Firefighter Anthony Calhoun died in 1992 when a train struck his vehicle as he responded to a wildland fire.

December 25, 2003 – 0930hrs
Kenneth J. "Kenny" Jeffery, Assistant Chief
Age 51, Career
Submarine Base Fire Department, Groton, Connecticut

Assistant Chief Jeffery reported for duty at 0730hrs. After roll call he worked out. At approximately 0908hrs, he was dispatched to an automatic fire alarm activation on the base. He proceeded to the fifth floor of the building to check on an alarm device. When he returned to the ground floor, he stated that he did not feel well.

When he arrived back at fire headquarters, he reported to coworkers that he was experiencing discomfort in his chest. He was treated by fire department EMT's and transported via fire department ambulance to the hospital. He was admitted to the hospital and treated for a heart attack. Within a few hours, he was transported by a medical helicopter to a cardiac care unit, where he died on December 31, 2003. The cause of death was listed as a ruptured aorta.

Assistant Chief Jeffery was the Fire Chief of the Windsor Locks Fire Department.

Windsor Locks Fire Department Web site - www.wlfd.com

December 29, 2003 – 1935hrs
Glyn Allen Taylor, Firefighter
Age 43, Volunteer
Jeff Davis Parish Fire District No. 3/Woodlawn Volunteer F.D., Louisiana

Firefighter Taylor was the driver of a brush truck returning to quarters from training. His apparatus was the last truck in a five-vehicle group of fire apparatus on the roadway.

A battery cover fell off of the lead truck and the five trucks pulled to the side of the roadway. It was dark at the time so the vehicle headlights were on. No warning lights were activated.

A number of firefighters dismounted their apparatus. One firefighter was detailed to pick up the cover and it was decided that the other four trucks would leave and wait for the fifth truck at a rest stop ahead. As firefighters prepared to board their apparatus, Firefighter Taylor stepped into the path of an oncoming car.

The left front bumper of the car struck firefighter Taylor and he was propelled over the top of the car. Firefighter Taylor fell into the roadway and was struck by a second vehicle.

Firefighters immediately went to assist Firefighter Taylor. He was transported to the hospital but died of his injuries. The cause of death was listed as multiple trauma.

Firefighter Taylor was not wearing any form of retroreflective materials and the apparatus headlights might have blinded the oncoming drivers.

Pre-2003 Incidents

October 15, 1987 – Time Unknown
Barry M. Bennett, Lieutenant
Age 33, Career
Cambridge Fire Department, Massachusetts

Firefighter Bennett suffered an infectious disease exposure while delivering emergency medical service on October 15, 1987. The needle stick exposure was immediately reported and a course of medical treatment and observation was begun. A more aggressive treatment was initiated when Lieutenant Bennett tested positively for Hepatitis-C.

Firefighter Bennett continued to work and was promoted to Lieutenant in 1998. By early 2003, Lieutenant Bennett became too ill to continue his duties. He received a transplanted liver in May of 2003.

Despite all efforts, Lieutenant Bennett died on November 2, 2003. He was 49 at the time of his death.

Cambridge Fire Department Web site - www.cambridgefire.org

July, 1996 – Time Unknown
Harry Zilkan, Assistant Chief
Age 49, Volunteer
Newberry Springs Fire Department, California

Assistant Chief Zilkan suffered a major heart attack while directing operations at a residential structure fire. He was revived enroute to the hospital. He heart was severely damaged.

A year after his heart attack, Assistant Chief Zilkan received a heart transplant. He was able to return to limited duties for a time but the transplanted heart began to fail, leading to his death on September 10, 2003.

June 2, 2001 – 2250hrs
Willard Burnis Paul, Firefighter
Age 64, Volunteer
Plainview Volunteer Fire Department, Louisiana

Firefighter Paul and the members of his fire department were performing stand-by duty at a local speedway. During the race, one of the cars lost control and hit the wall. An object, likely a fire extinguisher, that was either on top of the wall or near it was propelled by the force of the impact and struck Firefighter Paul in the head.

Members of his fire department came to his aid immediately. He was transported to the hospital and pronounced dead later that night. The cause of death was listed as an open head injury.

September 16, 2001 – 1045hrs
William Douglas "Doug" Thomas, Second Assistant Chief
Age 39, Volunteer
Kent Island Volunteer Fire Department, Maryland

Second Assistant Chief Thomas was severely injured in a fire apparatus crash on September 16, 2001. Second Assistant Chief Thomas was the driver of an engine that left the right side of the roadway, crashed into two utility poles, came back on the road, and overturned. The crash occurred during a response to a boat fire.

Second Assistant Chief Thomas was paralyzed from the chest down as a result of the crash. He was not able to return to duty. His health degenerated to the point that life support was removed on July 9, 2003. He died a short time later.